Photonic and Optoelectronic Devices and Systems, Third Edition

Guest Editor

Muhammad Ali Butt

MDPI

Basel • Beijing • Wuhan • Barcelona • Belgrade • Novi Sad • Cluj • Manchester

Guest Editor
Muhammad Ali Butt
Institute of Microelectronics
and Optoelectronics
Warsaw University of
Technology
Warsaw
Poland

Editorial Office
MDPI AG
Grosspeteranlage 5
4052 Basel, Switzerland

This is a reprint of the Special Issue, published open access by the journal *Micromachines* (ISSN 2072-666X), freely accessible at: https://www.mdpi.com/journal/micromachines/special_issues/RSLTHDEM7L.

For citation purposes, cite each article independently as indicated on the article page online and as indicated below:

Lastname, A.A.; Lastname, B.B. Article Title. *Journal Name* **Year**, *Volume Number*, Page Range.

ISBN 978-3-7258-6272-6 (Hbk)
ISBN 978-3-7258-6273-3 (PDF)
https://doi.org/10.3390/books978-3-7258-6273-3

Contents

About the Editor

Muhammad Ali Butt

Muhammad Ali Butt received his Ph.D. in Materials Science from Universitat Rovira i Virgili, Spain, in 2015. He is currently serving as an Assistant Professor at the Warsaw University of Technology, Poland. In 2018, he held the position of Research Assistant Professor at Nicolaus Copernicus University, Poland. His international research experience includes a stay at the Optoelectronics Research Centre (ORC), University of Southampton, United Kingdom, in 2013; a recent research visit to the Universitat Politècnica de València (UPV), Spain, in 2024; and a visit to Vrije Universiteit Amsterdam, Netherlands, in 2025. Dr. Butt has authored over 200 peer-reviewed scientific publications and book chapters. His research contributions have been recognized globally, and he was listed among the World's Top 2% Most Cited Researchers in 2021-2025. He also serves as an Editorial Board Member of *Scientific Reports*. His research interests encompass integrated photonics, metamaterials, numerical modelling, light–matter interaction, and plasmonic sensors.

Preface

This Reprint brings together a collection of contemporary research devoted to photonic and optoelectronic devices and systems. Its scope spans advances in micro- and nanoscale technologies, including waveguide engineering, detector architectures, display optimization, metasurface-enabled sensing, and high-speed optoelectronic components. The aim is to present readers with an integrated perspective on emerging methods, device concepts, and simulation-driven analysis that are shaping next-generation optical and electronic platforms.

The motivation for assembling this Reprint stems from the rapid evolution of photonics and semiconductor technologies, which continue to influence modern scientific development across sensing, communications, imaging, and display engineering. By gathering both review-level insights and specialized studies, this Reprint provides a coherent resource for understanding the technical progress highlighted in the contributing works. It is intended primarily for researchers, graduate students, and engineers engaged in photonic integration, semiconductor device design, optical sensing, and related applied sciences. I hope this Reprint will serve as a valuable reference for ongoing innovations in these dynamic fields.

Muhammad Ali Butt
Guest Editor

micromachines

Editorial

Photonic and Optoelectronic Devices and Systems, Third Edition

Muhammad A. Butt

Institute of Microelectronics and Optoelectronics, Warsaw University of Technology, Koszykowa 75, 00-662 Warsaw, Poland; ali.butt@pw.edu.pl

The rapid evolution of micro- and nanoscale technologies continues to shape modern scientific development across photonics, semiconductor devices, sensing, displays, and optical systems [1–4]. This Special Issue of *Micromachines* brings together ten contributions that collectively demonstrate the breadth and vitality of current research in these areas. Among them is one comprehensive review on transverse mode adaptive control based on photonic lanterns, accompanied by nine original research articles that address important advances in optical waveguides, high-purity and three-dimensional detectors, metasurface-based photodetection, electrowetting display optimization, avalanche photodiodes, and compact lens design for augmented reality. This editorial highlights the core contributions of each paper and reflects on the collective impact of the Issue.

The review article by Lu et al. provides a timely and forward-looking exploration of transverse mode adaptive control using photonic lantern technology [contribution 1]. The authors clearly articulate the limitations of existing mode control techniques and present photonic lanterns as a powerful pathway for achieving high-purity and rapidly reconfigurable multimode outputs in fiber systems. Their discussion spans device structures, fabrication approaches, mode evolution mechanisms, and integrating adaptive algorithms. The review underscores that photonic lantern-based adaptive systems offer compactness, fast response, low loss, and excellent scalability. The authors also identify impactful application fields that include suppressing transverse mode instability, mode division multiplexing, particle manipulation, and advanced spectral measurement. This study serves as an essential reference for researchers engaged in high-power fiber laser engineering, photonic integration, and adaptive optical control.

The nine original research articles in this Issue present significant advances across diverse domains. The study by Butt et al. investigates silicon nitride (SiN) ridge waveguides for enhanced refractive index sensing, combining numerical modeling with experimental validation [contribution 2]. Through a precise analysis of the effective refractive index, evanescent field ratio, and propagation losses, the authors demonstrate how waveguide geometry and wavelength selection can be leveraged to achieve improved sensitivity. Their experiments using a racetrack ring resonator reveal a marked increase in sensitivity at longer wavelengths, offering meaningful insights for the design of highly responsive photonic sensors. This study led to the further development of more advanced refractive index-sensing devices [5].

High-purity germanium drift detectors are examined in the study by Wang et al., who introduce an innovative design to address capacitance challenges in conventional detectors [contribution 3]. By using a small-area central cathode surrounded by concentric anode rings linked by a resistive chain, the detector achieves a capacitance determined solely by the collecting electrode rather than the overall device size. TCAD simulations confirm smooth electric potentials, directed lateral drift channels, and strong collection performance under heavy ion irradiation. This research provides a promising new direction for radiation detection instrumentation used in dark matter searches, high-energy physics, and space-based sensing.

Yoo et al. present a significant improvement in inverted organic light-emitting devices using lithium-doped magnesium zinc oxide nanoparticles as the electron injection layer [contribution 4]. The nanoparticles exhibit reduced particle size, modified energy band positions, and enhanced surface and electrical characteristics as lithium content increases. OLED devices fabricated with these modified layers achieve markedly higher external quantum efficiency through improved charge balance. Notably, the maximum efficiency reaches more than 20%, demonstrating the effectiveness of lithium doping as a strategy for tuning electron injection properties.

Three-dimensional (3D) silicon detector technology is the focus of two contributions. The first, by Zhu et al., proposes a polysilicon fill-strengthened etch through a 3D trench electrode detector that resolves the issue of non-uniform fields and dead zones in traditional trench designs [contribution 5]. Using double-sided etching and filling, the authors achieve a fully penetrating trench structure that improves electric field symmetry and reduces depletion voltage. Their simulations show extremely low leakage current, low capacitance, and efficient charge collection, indicating strong potential for demanding applications in nuclear radiation detection and particle physics.

The second detector-related paper by Liu et al. introduces a novel through-type semi-spherical electrode detector built on an SOI substrate [contribution 6]. The unique geometry equalizes horizontal and vertical dimensions to create a quasi-spherical electric field distribution. The result is a device with remarkably small capacitance, low depletion voltage, and excellent isolation between units, which is especially valuable for high-resolution X ray photon counting. This design highlights the valuable synergy between structural creativity and device physics modeling.

Cheng et al. contribute a design for a long-wave infrared fully polarized HgCdTe photodetector that employs a silicon metasurface to enable both circular and linear polarization resolution within a single compact platform [contribution 7]. The study integrates COMSOL (version 5.6) simulations to reveal how chiral silicon nanostructures and linear gratings work together to decode Stokes parameters without multilayer stacking. The demonstrated extinction ratio of thirty decibels for circular polarization indicates impressive performance. This effort expands the possibilities for compact integrated polarization-sensitive detection in spectroscopy, thermal imaging, and quantum information processing.

Complementing these studies, Yang et al. offer an electrowetting display driving strategy designed to reduce luminance instability caused by overdriving pulses [contribution 8]. The proposed three-stage pulse incorporates an overdriving phase for rapid contraction, a linear switching phase to suppress luminance glitches, and a final driving phase. Both simulations and experiments confirm that this method eliminates luminance instability while simultaneously increasing brightness and decreasing response time. This paper presents an important contribution to the advancement of high-quality reflective display technologies.

The article by Cheng et al. presents a three-terminal silicon germanium avalanche photodiode with a breakdown voltage of only 6.8 volts and an impressive gain bandwidth product of more than 1300 gigahertz [contribution 9]. The device employs a separate absorption and multiplication structure and integrates a silicon nitride waveguide for lateral optical coupling. Simulations show excellent responsivity, high-speed modulation capability, and clean eye diagrams at 100 and 200 gigabits-per-second data rates. This study holds strong relevance for silicon photonics-based receivers in advanced communication systems.

Finally, the compact collimating lens system introduced by Sun et al. demonstrates a practical solution for generating a fifty-degree field of view in lightguide-based augmented reality glasses [contribution 10]. The design uses four plastic aspherical lenses to achieve high optical quality, low distortion, and a very small form factor compatible with a 0.32-inch microdisplay. With a total volume below one cubic centimeter, this optical module advances the goal of lightweight and unobtrusive AR eyewear.

Together, these ten papers offer a valuable snapshot of progress across micro- and nanoscale device engineering [6,7]. They reveal a consistent push toward compactness, efficiency, precision, and integrability across materials and platforms. This Special Issue demonstrates the increasing convergence of photonics, semiconductor technology, sensing, and display innovation, and we hope it will inspire continued research that bridges these diverse yet synergistic fields.

Funding: This research received no external funding.

Acknowledgments: The Guest Editor acknowledges the support of all the authors in the completion of this Special Issue.

Conflicts of Interest: The authors declare no conflicts of interest.

List of Contributions:

1. Lu, Y.; Jiang, Z.; Chen, Z.; Sun, Z.; Liu, T. A Review of Transverse Mode Adaptive Control Based on Photonic Lanterns. *Micromachines* **2025**, *16*, 1347. https://doi.org/10.3390/mi16121347.
2. Butt, M.A.; Kozlowski, L.; Słowikowski, M.; Juchniewicz, M.; Drecka, D.; Filipiak, M.; Golas, M.; Stonio, B.; Dudek, M.; Piramidowicz, R. Investigation of Modal Characteristics of Silicon Nitride Ridge Waveguides for Enhanced Refractive Index Sensing. *Micromachines* **2025**, *16*, 119. https://doi.org/10.3390/mi16020119.
3. Wang, M.; Li, Z.; Xiong, B.; Xiao, Y. TCAD Simulation Study of Electrical Performance of a Novel High-Purity Germanium Drift Detector. *Micromachines* **2025**, *16*, 229. https://doi.org/10.3390/mi16020229.
4. Yoo, H.-J.; Kim, G.-E.; Park, C.-J.; Lee, S.-B.; Kim, S.-Y.; Moon, D.-G. Highly Efficient Inverted Organic Light-Emitting Devices with Li-Doped MgZnO Nanoparticle Electron Injection Layer. *Micromachines* **2025**, *16*, 617. https://doi.org/10.3390/mi16060617.
5. Zhu, X.; Li, Z.; Liu, Z.; Long, T.; Zhao, J.; Li, X.; Liu, M.; Wang, M. A Novel Polysilicon-Fill-Strengthened Etch-Through 3D Trench Electrode Detector: Fabrication Methods and Electrical Property Simulations. *Micromachines* **2025**, *16*, 912. https://doi.org/10.3390/mi16080912.
6. Liu, Z.; Long, T.; Li, Z.; Zhu, X.; Zhao, J.; Li, X.; Liu, M.; Wang, M. Development, Design, and Electrical Performance Simulation of Novel Through-Type 3D Semi Spherical Electrode Detector Based on SOI Substrate. *Micromachines* **2025**, *16*, 1006. https://doi.org/10.3390/mi16091006.
7. Cheng, B.; Wang, X.; Zou, Y.; Song, G.; Zhai, K.; Wang, X. Design of Long-Wave Fully Polarized HgCdTe Photodetector Based on Silicon Metasurface. *Micromachines* **2025**, *16*, 937. https://doi.org/10.3390/mi16080937.
8. Yang, Y.; Yi, Z.; Xu, W.; Wang, J.; Lu, Q.; Liu, Q.; Liu, L.; Chi, F. Optimization of Overdriving Pulse for Luminance Stability of Electrowetting Displays. *Micromachines* **2025**, *16*, 1085. https://doi.org/10.3390/mi16101085.
9. Cheng, C.; Xue, J.; Yu, X.; Mu, J.; Wang, B. A Three-Terminal Si-Ge Avalanche Photodiode with a Breakdown Voltage of 6.8 V and a Gain Bandwidth Product of 1377 GHz. *Micromachines* **2025**, *16*, 1222. https://doi.org/10.3390/mi16111222.
10. Sun, W.-S.; Su, Y.-L.; Hsu, Y.-S.; Tien, C.-L.; Cheng, N.-J.; Sun, C.-C. Compact Design of a 50° Field of View Collimating Lens for Lightguide-Based Augmented Reality Glasses. *Micromachines* **2025**, *16*, 1234. https://doi.org/10.3390/mi16111234.

References

1. Song, R.; Cho, S.; Khan, S.; Park, I.; Gao, W. Lighting the Path to Precision Healthcare: Advances and Applications of Wearable Photonic Sensors. *Adv. Mater.* **2025**, 2419161. [CrossRef] [PubMed]
2. Striakhilev, D.; Nathan, A.; Vygranenko, Y.; Servati, P.; Lee, C.-H.; Sazonov, A. Amorphous Silicon Display Backplanes on Plastic Substrates. *J. Disp. Technol.* **2006**, *2*, 364–371. [CrossRef]
3. Alimi, I.A.; Monteiro, P.P. Revolutionizing Free-Space Optics: A Survey of Enabling Technologies, Challenges, Trends, and Prospects of Beyond 5G Free-Space Optical (FSO) Communication Systems. *Sensors* **2024**, *24*, 8036. [CrossRef]
4. Cho, Y.-C.; Gupta, K.; Moon, J.; Choi, W. Experimental Realization of an All-Optical Computing System Using a Coherent Nonlinear Scattering Medium. *Opt. Commun.* **2025**, *589*, 132002. [CrossRef]
5. Butt, M.A.; Słowikowski, M.; Drecka, D.; Jarosik, M.; Kozłowski, Ł.; Piramidowicz, R. Design and Implementation of Silicon Nitride Resonant Cavities for Refractive Index Detection. *Opt. Lasers Eng.* **2026**, *196*, 109444. [CrossRef]
6. Li, J.; Yan, J.; Jiang, L.; Yu, J.; Guo, H.; Qu, L. Nanoscale Multi-Beam Lithography of Photonic Crystals with Ultrafast Laser. *Light Sci. Appl.* **2023**, *12*, 164. [CrossRef] [PubMed]
7. Cueff, S.; Poon, J.; Thourhout, D.V.; Vivien, L. Hybrid Photonics: Integration, Design and Devices: Feature Issue Introduction. *Opt. Mater. Express OME* **2024**, *14*, 1456–1458. [CrossRef]

micromachines

MDPI

Article

Compact Design of a 50° Field of View Collimating Lens for Lightguide-Based Augmented Reality Glasses

Wen-Shing Sun [1], Yi-Lun Su [1], Ying-Shun Hsu [1], Chuen-Lin Tien [2,*], Nai-Jen Cheng [3] and Ching-Cherng Sun [1]

[1] Department of Optics and Photonics, National Central University, Chungli 32001, Taiwan; wssun@dop.ncu.edu.tw (W.-S.S.); a0357841@gmail.com (Y.-L.S.); dmcandymurray@yahoo.com.tw (Y.-S.H.); ccsun65298@gmail.com (C.-C.S.)
[2] Department of Electrical Engineering, Feng Chia University, Taichung 40724, Taiwan
[3] Institute of Photonics Engineering, National Kaohsiung University of Science and Technology, Kaohsiung 807618, Taiwan; njcheng@nkust.edu.tw
* Correspondence: cltien@fcu.edu.tw; Tel.: +886-4-24517250 (ext. 3809)

Abstract: Designing a compact collimating lens system for augmented reality (AR) applications presents significant optical challenges. This paper presents a compact, 50-degree field-of-view collimating lens system explicitly designed for lightguide-based AR glasses. The compact collimating lens is designed for a 0.32-inch microdisplay and consists of four plastic aspherical lenses. The optical design results in a collimating lens with a F-number of 2.17 and an entrance pupil diameter of 4 mm. Optical distortion is less than 0.29%, and the modulation transfer function (MTF) is greater than 0.23 at 250 cycles/mm. The overall lens diameter, including the lens barrel, measures 10.16 mm, while the lens length is 11.48 mm. The lens volume is 0.93 cm^3, and its mass is 1.08 g. Compared to existing collimator designs, this approach significantly improves the trade-off between field of view, optical quality, and device miniaturization. The proposed design supports integration with 0.32-inch microdisplays, making it a practical and manufacturable solution for next-generation AR eyewear. This paper presents innovative contributions to the optical design of AR glasses, demonstrating considerable potential in reducing size and weight, and optimizing optical performance.

Keywords: compact collimating lens; augmented reality; microdisplay; modulation transfer function

1. Introduction

In recent years, augmented reality (AR) lightguide displays have gained increasing attention due to their potential to deliver lightweight, wide field-of-view (FOV), and high-quality visual experiences in wearable systems. Among the various optical coupling and image-forming technologies, volume holographic optical elements (VHOEs) have emerged as a promising solution for achieving high diffraction efficiency, precise wavefront control, and compact system integration [1–4]. The collimating lens is a crucial optical component of head-mounted displays (HMDs), responsible for projecting images from the display to the human eye. Extensive research has been devoted to enhancing the design, simulation, and fabrication of VHOE-based lightguides. For example, Tsai [5] integrates Kogelnik's coupled-wave theory with ray tracing modeling to optimize lightguide combiners, improving efficiency, reducing module volume, and enhancing image quality. Similarly, Lee [6] embeds coupled-wave theory into ray-tracing-based simulations, enabling accurate modeling of VHOE diffraction and image performance in AR systems. Efforts toward structural optimization have also been reported.

The resolution of the collimating lens can significantly affect the overall optical quality of the system, including light source utilization, illumination uniformity, color reproduction, and image performance. Hua [7] present the design of an ultralight and compact collimating lens for a HMDs, utilizing a 1.35-inch microdisplay with a resolution of 640 × 480 pixels, providing a 52.4 degrees FOV, an F-number of 2.92, an entrance pupil diameter of 12 mm, and an effective focal length of 35 mm, with a total volume of 3.53 cm^3. In 2007, Hua [8] proposed a polarized head-mounted projection display (p-HMPD) and developed a compact optical system using a 1.3-inch microdisplay of 640 × 480 pixels, achieving 56 degrees FOV, an F-number of 3.2, an entrance pupil diameter of 10 mm, and an effective focal length of 31.5 mm, with a reduced lens volume of 3.15 cm^3. In 2008, Zhang and Hua [9] reported the design of a p-HMPD prototype employing a pair of high-resolution ferroelectric liquid-crystal-on-silicon (FLCOS) microdisplays, offering higher optical efficiency. The light engine used a 0.88-inch microdisplay at 1280 × 1024 pixels, providing a 55-degree FOV, an F-number of 2.16, an entrance pupil diameter of 10 mm, and an effective focal length of 21.6 mm, with a compact lens volume of approximately 2.66 cm^3. Pan [10] designed a collimating lens for a geometric waveguide display using a 0.62-inch microdisplay with 800 × 600 pixels, achieving a 30-degree FOV, an F-number of 3.66, an entrance pupil diameter of 8 mm, and an effective focal length of 29.3 mm, with a total lens volume of 12.45 cm^3. Tsai [11] proposed an optical see-through head-mounted display (OST-HMD) system in which a rotationally symmetric eyepiece with a wide 60-degree FOV was coupled with a dihedral corner reflector array (DCRA). The system employed a 0.7-inch microdisplay with a resolution of 1920 × 1080 pixels, an F-number of 2.58, an entrance pupil diameter of 6 mm, an effective focal length of 15.46 mm, and a lens volume of 9.76 cm^3. Wu [12] proposes a curved variable-period grating to replace conventional in-coupler and collimating lens, achieving about 39.3% thickness reduction, 70% system efficiency, and a 36.6 degrees FOV. Weng [13] proposed a waveguide display scheme based on the coupling-collimating system. This system incorporated a freeform liquid crystal diffractive optical element within the waveguide and a freeform element outside the waveguide, achieving a FOV of 35 degrees, an F-number of 2.5, an entrance pupil diameter of 9 mm, and an effective focal length of 22.5 mm.

Sun [14] designed a collimating lens with a 30-degree FOV for a 0.35-inch microdisplay with 1280 × 720 pixels. The entrance pupil diameter of the lens is 14 mm, an effective focal length of 16.443 mm with an F-number of 1.175, and the lens volume is 25.02 cm^3. In 2025, Sun [15] proposed a wide FOV of a 65-degree collimating lens design. The entrance pupil diameter of this system is 10 mm, the effective focal length is 20.458 mm, and the F-number is 2.046. It features a 1.03-inch microdisplay with a resolution of 2560 × 2560 pixels; the lens volume is 22.61 cm^3. This paper presents innovative contributions to the optical design of augmented reality glasses, demonstrating considerable potential in reducing size and weight, and optimizing optical performance. The final lens volume is 0.93 cm^3, and the mass is 1.08 g.

2. Optical Design Method

2.1. Subsection Microdisplay Specifications

The design of a collimating lens with a 50-degree FOV was a 0.32-inch microdisplay. The specifications are shown in Table 1. In the specifications, the pixel size was 8.1 μm × 8.1 μm. The effective area of 6.48 mm × 4.86 mm was calculated for the diagonal length of 8.1 mm, which could be used to define the paraxial image height of 4.05 mm.

Table 1. Microdisplay specifications.

Parameters	Specification
Pixel number	800 × 600
Pixel size	8.1 μm × 8.1 μm
Effective area	6.48 mm × 4.86 mm
Diagonal length of the display	8.1 mm (0.32 inch)

2.2. Five Reference Rays on the Entrance Pupil

In the CODE V optical design software, each field of view is associated with five reference rays, denoted as R1, R2, R3, R4, and R5, as illustrated in Figure 1. Among them, R1 represents the chief ray, which passes through the center of the entrance pupil and forms an angle θ with the optical axis, corresponding to the half FOV. The remaining rays—R2, R3, R4, and R5—are marginal rays that pass through the top edge (+Y), bottom edge (−Y), left edge (+X), and right edge (−X) of the entrance pupil, respectively.

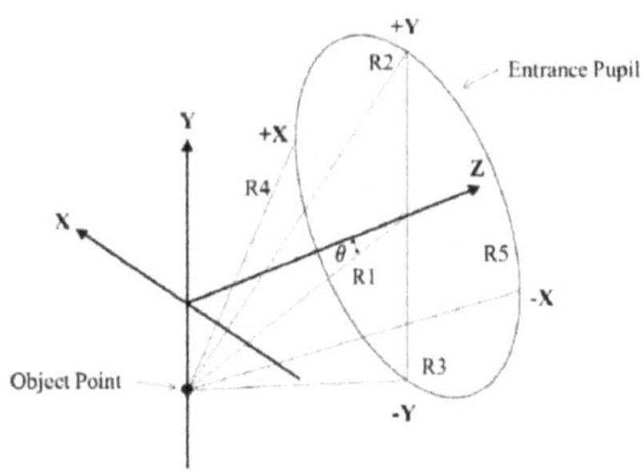

Figure 1. The five reference rays (R1, R2, R3, R4, and R5) defined on the entrance pupil.

2.3. Clear Aperture Calculation of the Lens

The maximum half FOV of 25° is defined as the full field (1.0 field) to define the clear aperture of the collimating lens system. To ensure uniform image quality, the 1.0 field is divided into 11 evenly spaced points for optimization, corresponding to 0.0, 0.1, 0.2, . . . , up to 1.0 field. In CODE V, the clear aperture of each lens surface is calculated by performing real ray tracing of five reference rays (R1 to R5) per field point, resulting in a total of 55 rays. The maximum footprint of these rays on each surface defines the required clear aperture of that surface. The collimating lens system in this study covers a FOV of 50°. It comprises four lens elements and one protective window, as illustrated in Figure 2. The optical axis is aligned with the Z-axis. Figure 2a shows the Y–Z cross-sectional view of the lens system and the ray traces for reference rays R1, R2, and R3 across all fields (a total of 33 rays). Figure 2b shows the X–Z plane, which includes reference rays R4 and R5 (a total of 22 rays). Hence, to fully determine the clear aperture of each lens surface, ray tracing of all 55 reference rays is required.

2.4. Relationship Among Half FOV, Image Height, Entrance Pupil Diameter, and Effective Focal Length

Figure 3 illustrates the relationships among half FOV, image height, effective focal length (EFL), and entrance pupil diameter. Points P and P′ denote the first and second principal points, respectively, while N and N′ denote the first and second nodal points. The symbols n and n′ represent the refractive indices of the object and image spaces, respectively. F′ denotes the rear focal point, h′ represents the image height, and STO indicates the position of the aperture stop. In this system, the aperture stop is located on the first surface of the lens, causing the entrance pupil to coincide with the aperture stop. EXP indicates the exit pupil position, while D_{en} and D_{ex} represent the diameters of the entrance and exit pupils, respectively. The symbol θ_{CA} refers to the chief ray angle in the image space. EFL is the lens's effective focal length, and BFL is the back focal length. Assuming that the object space half FOV is denoted by θ, and the object is located at infinity, then the chief ray, marginal ray, and the ray passing through the first principal point (P) will all make the same angle θ. If the object space and image space refractive indices (n and n′) are equal (typically air), then the first principal point P coincides with the first nodal point N, and the second principal point P′ coincides with the second nodal point N′. Due to the nodal point property, the incident angle at N equals the emergent angle at N′, thereby establishing the following relationship between image height h′ and the effective focal length EFL, as shown in Equation (1)

$$tan\theta = \frac{h'}{EFL}. \tag{1}$$

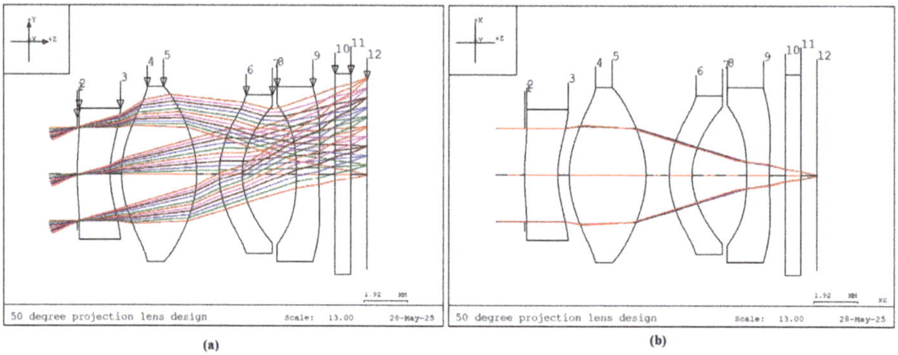

Figure 2. Ray trace diagrams of the 50° FOV collimating lens design. (**a**) Lens elements and reference rays in the Y–Z plane. (**b**) Lens elements and reference rays in the X–Z plane.

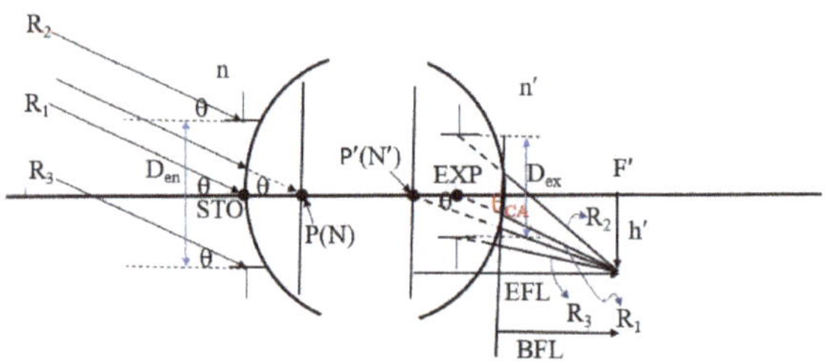

Figure 3. Geometric relationship among half field angle, image height, effective focal length, and entrance pupil diameter.

Furthermore, the relationship among the F-number, the entrance pupil diameter (Den), and the effective focal length (EFL) is given by Equation (2).

$$\text{F-number} = \frac{EFL}{D_{en}}. \tag{2}$$

The clear aperture of each lens surface decreases as the entrance pupil diameter (D_{en}), image height (h′), or half field angle (θ) decreases. Since the half FOV θ is fixed at 25°, Equation (1) implies that a smaller image height h′ corresponds to a shorter effective focal length (EFL). According to Equation (2), if the entrance pupil diameter D_{en} is fixed at 4 mm, a decrease in EFL also results in a lower F-number.

2.5. Entrance Pupil Position and Size

The aperture stop location and size determine the minimum clear aperture of a lens system. When the aperture stop is placed on the first surface or before the lens, the entrance pupil (ENP) coincides with the aperture stop. Its size is identical to that of the aperture stop, and directly determines the clear aperture of the first lens, which in turn influences the clear aperture of the second lens, as illustrated in Figure 4. As shown in Figure 4a, with the aperture stop on the first surface of lens, the minimum clear aperture occurs on the first lens. Figure 4b shows the aperture stop before the lens, where the entrance pupil coincides with the aperture stop. The ray height of reference ray R3 on the first surface of the lens is larger than the entrance pupil radius. If the aperture stop position moves toward the object space, the farther the distance the aperture stop is from the first surface of the lens, the greater the ray height of reference ray R3 on the first surface, and a larger clear aperture of the first lens, and increasing the overall lens volume. Figure 4c shows the entrance pupil behind the first surface, which no longer coincides with the aperture stop. The ray height of reference ray R2 on the first surface of the lens is larger than the entrance pupil radius. If the entrance pupil position moves toward the image space, the farther the entrance pupil is from the first surface, the greater the ray height of reference ray R2 on the first surface, and a larger clear aperture of the first lens, increasing the overall lens volume.

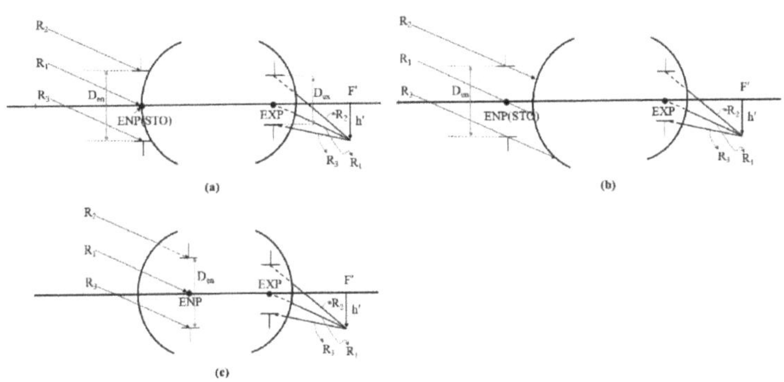

Figure 4. Relationship between the entrance pupil position and the clear aperture of the first lens: (**a**) entrance pupil located on the first surface; (**b**) entrance pupil located in front of the first surface; (**c**) entrance pupil located behind the first surface.

2.6. Exit Pupil Position and Size

The FOV 50-degree collimating lens design consists of four lenses. Since the aperture stop is located on the first lens, the first lens has the smallest clear aperture. However, because the third and fourth lenses are farther away from the first surface of the lens, the clear apertures of the third and fourth lenses tend to increase. We must suppress the factors that cause the clear apertures of the third and fourth lenses to become larger.

Figure 5 defines the relationship between the exit pupil position and size and the clear aperture of the rearmost surface of the lens. Distances are directional, with negative values to the left and positive values to the right. S is the distance from the first principal point (P) to the entrance pupil (ENP), and S' is the distance from the second principal point (P′) to the exit pupil (EXP). δ is the distance from the first surface of the lens to the first principal point, and δ' is the distance from the rearmost surface of the lens to the second principal point. Therefore, the distance from the first surface of the lens to the entrance pupil is

$S + \delta$, and the distance from the rearmost surface of the lens to the exit pupil is $S' + \delta'$. If the exit pupil is at the second principal point, the image space chief ray angle θ_{CA} equals the half FOV θ, $\theta_{CA} = 25°$. If the exit pupil position is to the right of the second principal point, $\theta_{CA} > 25°$; and if it is to the left of the second principal point, $\theta_{CA} < 25°$. As θ_{CA} decreases, the exit pupil becomes farther from the rearmost surface of the lens (excluding the flat glass), resulting in a larger clear aperture at the rearmost lens. If $\theta_{CA} = 0°$, the exit pupil position is at infinity, meaning the image space chief ray is incident perpendicularly on the image plane (image telecentric), and the clear aperture of the rearmost element is at its maximum. Similarly, as θ_{CA} increases ($\theta_{CA} > 25°$), the exit pupil becomes closer to the rearmost surface of the lens, resulting in a smaller clear aperture. If the exit pupil is at the rearmost surface of the lens, the clear aperture of the rearmost lens is at its minimum, calculated as the exit pupil diameter (D_{ex}). Image height h' and exit pupil diameter are also important factors influencing the clear aperture of the rearmost lens.

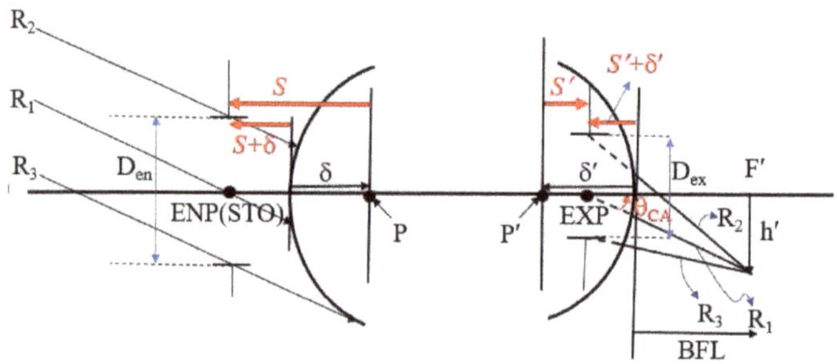

Figure 5. Relationship between the exit pupil position and the clear aperture of the last lens surface.

The exit pupil distance S' can be calculated from the relationship between S and S' and the EFL of the lens as shown in Equation (3) [16]:

$$\frac{1}{S'} = \frac{1}{S} + \frac{1}{EFL}. \tag{3}$$

The image space chief ray angle θ_{CA}, is determined according to Equation (4).

$$\theta_{CA} = \tan^{-1}\left(\frac{h'}{BFL - S' - \delta'}\right). \tag{4}$$

The exit pupil diameter (D_{ex}) is obtained through the relationship between the lens lateral magnification (M_T) and the entrance pupil diameter (D_{en}), as given in Equation (5) [16].

$$M_T = \frac{S'}{S} = \frac{D_{ex}}{D_{en}}. \tag{5}$$

2.7. Relationship Between Effective Focal Length and Angular Magnification

The angular magnification (M_P) of the magnifying glass [17] can be obtained

$$M_P = -\frac{d_o}{f'} \tag{6}$$

In the design described in this paper, the collimating lens function like a magnifying glass. As a result, the image is magnified after passing through the collimating lens. Let d_o be the distance of distinct vision, $d_o = -250$ mm, f' denote the image space focal length of the collimating lens, where $f' = 8.68$ mm. Let M_P represent the angular magnification. $M_P = 28.8°$ The effective area of the microdisplay is 6.48 mm × 4.86 mm, and then the horizontal length and the vertical length of the virtual image when the eyes can see from the collimating lens at a distance of 250 mm can be obtained by 186.624 mm × 139.968 mm [15].

3. Design Result

This section may be divided by subheadings. It should provide a concise and precise description of the design results, their interpretation, as well as the design conclusions that can be drawn.

3.1. Lens Specifications and Lens Data

The specifications of the collimating lens are shown in Table 2, with the aperture stop location on the first surface. The entrance pupil diameter is 4 mm, and the lens has a diagonal FOV of 50 degrees. A horizontal FOV of 40.94 degrees is required. The F-number is 2.17, the focal length is 8.68 mm, and the image height is 4.05 mm. The lens resolution is set at 62 cycles/mm, which corresponds to an angular resolution requirement of 20 pixels per degree (PPD). The angular magnification of the collimating lens is 28.8, and the design specifications are summarized in Table 2.

Table 2. The specifications of the collimating lens.

Parameters	Specifications
Entrance pupil	4 mm
Diagonal FOV	50°
Horizontal FOV	40.94°
Vertical FOV	31.28°
Lens clear aperture maximum	≤6.8 mm
Focal length	8.68 mm
F-number	2.17
Image height	4.05 mm
Protective glass material	BSC1(HOYA)
Protective glass thickness	0.7 mm
MTF (62 cycles/mm)	≥0.7
Angular resolution (pixel per degree)	20 PPD
Angular magnification	28.802

The lens layout of the ultra-thin collimating lens with a FOV of 50° is shown in Figure 6, and the corresponding lens data and aspherical surface data are listed in Tables 3 and 4, respectively. Surface Number is the order of the surface numbers of the collimating lens. K represents the conic constant. A, B, C, and D represent the coefficients of the 4th, 6th, 8th, and 10th order terms of the aspheric surface. The K value is defined as: $K > 0$ represents an elliptical surface with the minor axis on the optical axis, $K = 0$ represents a spherical surface, $-1 < K < 0$ represents an elliptical surface with the major axis on the optical axis, $K = -1$ represents a paraboloid, and $K < -1$ represents a hyperboloid. The system comprises four plastic aspheric lenses and one protective cover glass for the microdisplay.

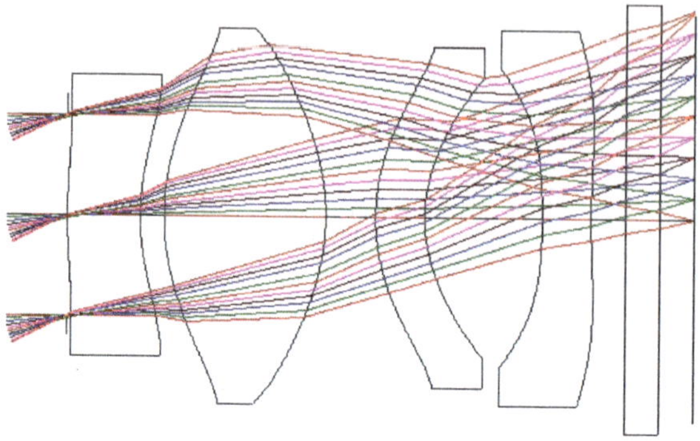

Figure 6. Optical layout of the 50° FOV collimating lens design.

Table 3. Lens design data of the collimating lens.

Surface Number	Surface Type	Radius (mm)	Thickness (mm)	Glass	Clear Aperture (mm)
Object	Sphere	Infinity	Infinity		
Stop	Sphere	Infinity	0		4.000
2	Asphere	10.6751	1.483	'OKP-1_30′	4.083
3	Asphere	4.4966	0.500		5.085
4	Asphere	4.3777	3.368	PMMA_SPECAL	6.378
5	Asphere	−4.3101	1.000		6.798
6	Asphere	3.8496	1.000	'OKP-1_30′	6.145
7	Asphere	2.9794	2.400		5.604
8	Asphere	−13.7379	1.000	'OKP-1_30′	5.719
9	Asphere	15.2466	0.700		6.774
10	Sphere	Infinity	0.700	BSC1_HOYA	7.438
11	Sphere	Infinity	0.700		7.776
12	Image	Infinity	0.000		8.307

Table 4. Aspherical surface data for 50-degree FOV collimating lens.

Surface Number	K	A	B	C	D
2	−120	0.335966×10^{-2}	-0.213406×10^{-2}	0.302968×10^{-3}	-0.192055×10^{-4}
3	−5.161698	-0.148158×10^{-2}	-0.606542×10^{-3}	0.717905×10^{-4}	-0.338550×10^{-5}
4	−0.800282	-0.149620×10^{-2}	-0.182126×10^{-3}	0.211560×10^{-4}	-0.100896×10^{-5}
5	−4.456061	-0.298062×10^{-2}	0.193746×10^{-3}	-0.382837×10^{-5}	-0.339511×10^{-6}
6	−0.966918	-0.721381×10^{-3}	-0.108456×10^{-3}	0.995474×10^{-5}	-0.119197×10^{-5}
7	−2.644408	0.454588×10^{-2}	-0.288402×10^{-3}	0.113949×10^{-4}	0.201825×10^{-6}
8	−74.465683	-0.133534×10^{-1}	0.113535×10^{-2}	-0.135322×10^{-3}	0.838188×10^{-5}
9	−15.581008	-0.702624×10^{-2}	0.481560×10^{-3}	-0.267416×10^{-4}	0.797422×10^{-6}

3.2. Analysis of Collimating Lens Image Quality

The image quality analysis of the collimating lens includes evaluation of the Modulation Transfer Function (MTF), optical distortion, lateral color, and relative illumination. Figure 7 shows the MTF curves of the lens, where the horizontal axis represents spatial frequency. Given that the pixel size of the microdisplay is 8.1 μm, the Nyquist frequency is determined to be 62 cycles/mm. At this frequency, the MTF value is at least 0.711. Since the resolution limit is constrained by the pixel size of the microdisplay (i.e., 62 cycles/mm), this defines the upper bound of the system performance when the display is considered. However, if the MTF performance of the collimating lens is evaluated independently—disregarding the pixel limitations of the microdisplay—and the same image plane position and image area are assumed, the spatial frequency can be extended up to 250 cycles/mm. At this frequency, the MTF value remains above 0.232, as shown in Figure 8, corresponding to an angular resolution of up to 79 pixels per degree (PPD).

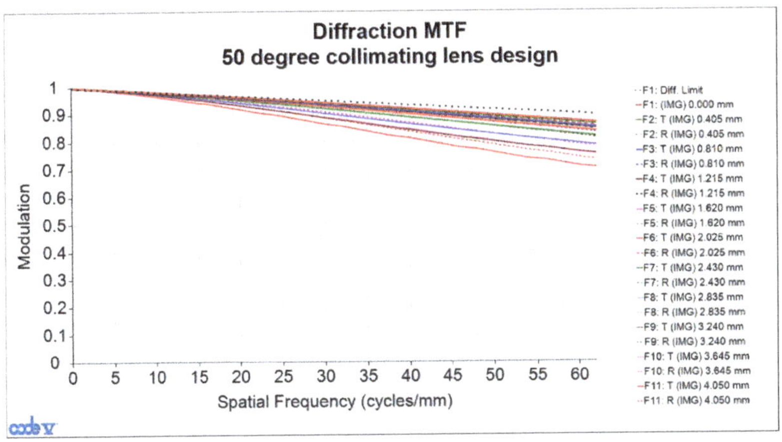

Figure 7. Modulation Transfer Function (MTF) curve of the FOV 50° collimating lens design at a spatial frequency of 62 cycles/mm.

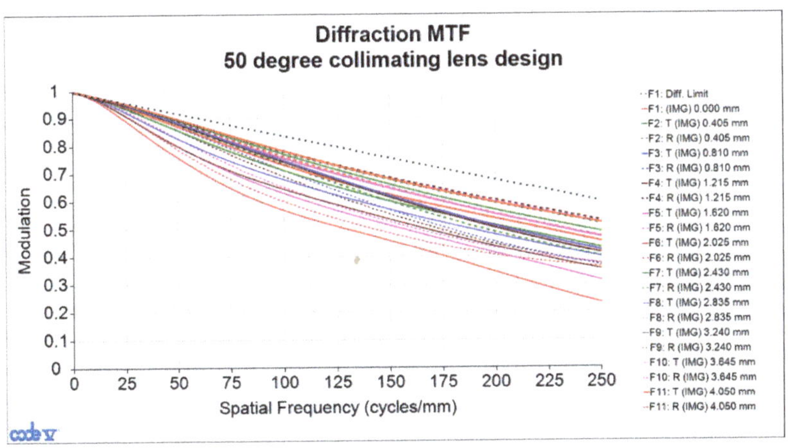

Figure 8. Modulation Transfer Function (MTF) curve of the FOV 50° collimating lens design at a spatial frequency of 250 cycles/mm.

Figure 9a shows the optical distortion plot for the collimating lens design. The maximum optical distortion is 0.2128%, occurring at the 0.8 field position. The TV distortion, defined as the absolute difference between the maximum and minimum optical distortion values from the 0.6 to 1.0 field region, is 0.0195%. Figure 9b presents the corresponding distortion grid chart. The horizontal and vertical axes represent the image heights in the horizontal and vertical directions. The black curve illustrates the variation in paraxial image height, while the red curve represents the actual image height. The collimator design in this study exhibits minimal TV distortion (0.0195%), resulting in a straight red curve without distortion. In addition, the collimator design in this article also exhibits minimal optical distortion (0.2128%), allowing the red curve to overlap with the black curve, making the black curve effectively invisible.

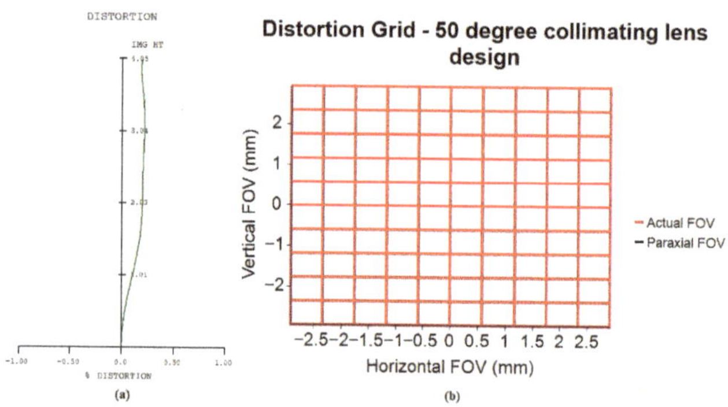

Figure 9. Optical distortion curve for 50-degree FOV collimating lens. (**a**) the optical distortion plot; (**b**) the corresponding distortion grid chart.

Lateral chromatic aberration is defined as the difference in image height on the imaging plane when tracing real light rays of different wavelengths. The lens design presented in this paper is based on three wavelengths: a short wavelength (F-line, 0.4861 μm), a reference wavelength (d-line, 0.5876 μm), and a long wavelength (C-line, 0.6563 μm). The design values for lateral color are shown in Figure 10. In this figure, the horizontal axis represents lateral color in mm, and the vertical axis represents the half-field angle in degrees. The red curve represents the lateral color values for the short and long wavelengths across different field angles. The maximum value of the red curve occurs at a half FOV of 25°, with a value of 0.742 μm. The green curve represents the lateral color values for the short and reference wavelengths. The maximum value of the green curve occurs at a half FOV of 25°, with a value of 1.426 μm. Lateral chromatic aberration must be smaller than the microdisplay pixel size of 8.1 μm; otherwise, the lens will be unable to resolve it, and lateral chromatic aberration will occur easily. The lateral chromatic aberration of the collimator designed in this paper is much smaller than the microdisplay pixel size of 8.1 μm, so the lateral chromatic aberration value can be ignored.

Finally, the relative illumination curve is shown in Figure 11. The red curve shows the relative illumination values at different half-field angles. In this design, the relative illumination is greater than 65%.

3.3. Lens Manufacturing and Tolerance Analysis

The tolerances [18] include the radius of curvature (DLF), the cylinder irregularity (CYD, CYN), the thickness (DLT), the refractive index (DLN), the V-number (DLV), the wedge (TRX, TRY), the tilt (BTX, BTY), and the displacement (DSX, DSY). The range of tolerance parameters set for this lens design is shown in Table 5.

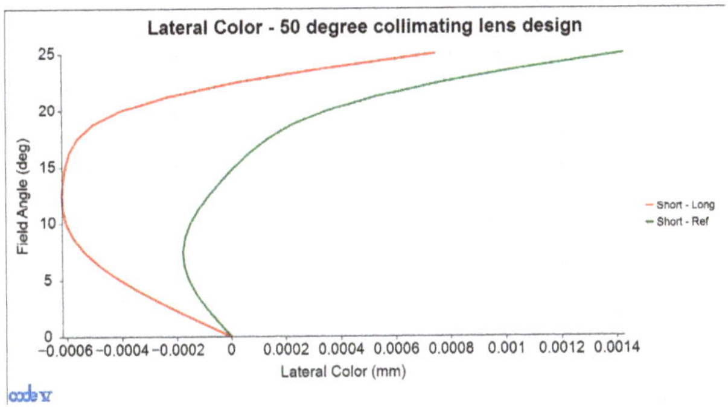

Figure 10. Lateral color curve for 50-degree FOV collimating lens.

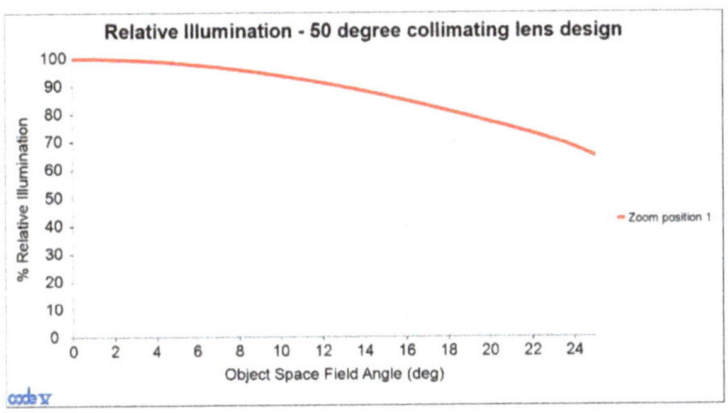

Figure 11. Relative illumination curve for 50-degree FOV collimating lens.

Table 5. Tolerance parameter range.

Type	Minimum	Maximum	Increment
DLF (fringe)	1	5	0.5
DLT (mm)	0.003	0.02	0.005
DLN	0.0001	0.0005	0.0001
DLV	0.001	0.005	0.001
CYD (fringe)	1	1.5	0.1
CYN (fringe)	1	1.5	0.1
TRX (arcmin)	1	3	0.1
TRY (arcmin)	1	3	0.1
BTY (arcmin)	1	3	0.1
BTX (arcmin)	1	3	0.1
DSX (mm)	0.002	0.01	0.001
DSY (mm)	0.002	0.01	0.001

The cumulative distribution probability curve for tolerance analysis is found by combining the effects of all tolerance items. The vertical axis represents the cumulative distribution probability, and the horizontal axis represents the MTF. Tan stands for tangential direction, and rad stands for sagittal direction. At a cumulative probability of 97.7%, the lowest MTF value at a spatial frequency of 62 cycles/mm for the tolerances at the 1.00 field tan position is 0.3023, as shown in Figure 12. Except for the 0-degree field of view, the MTF curve is affected by astigmatism in other fields of view, and its tangential and sagittal MTF curves will be different. The maximum paraxial image height of 4.05 mm is defined as the 1.0 field of the collimating lens system. To ensure uniform image quality, the 1.0 field is divided into 11 evenly spaced points, corresponding to 0.0, 0.1, 0.2, ..., up to 1.0 field. F1, F2, ..., F11 are represented by tangential fields of view of 0.0, 0.1, 0.2, ..., 1.0, respectively; and F12, F13, ..., F21 are represented by radial fields of view of 0.1, 0.2, ..., 1.0, respectively. In the tolerance analysis, we analyze each field of view, including tangential and radial values. We perform tolerance analysis on F1 to F21 in various directions across different fields of view, using different colors to represent each.

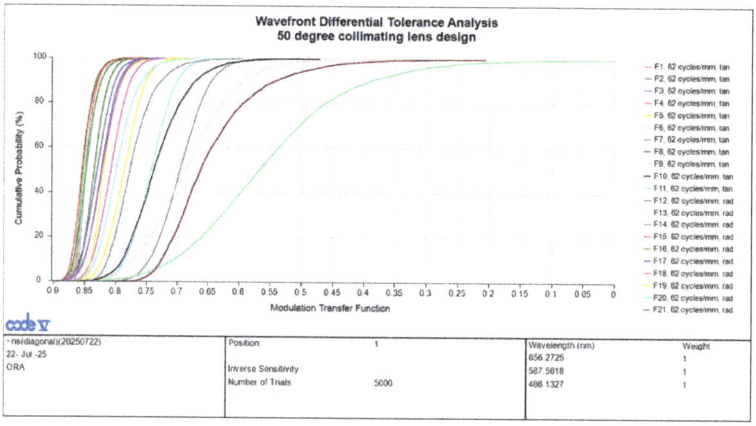

Figure 12. Tolerance analysis curve for 50-degree FOV collimating lens.

The values of the design MTF and design plus tolerance MTF for both the tangential and radial directions at a spatial frequency of 62 cycles/mm across all fields are shown in Table 6.

Table 6. Design and tolerance MTF at spatial frequency 62 cycles/mm.

Field	Spatial Frequency (cycles/mm)	Azimuth	Design MTF	Design Plus Tolerance MTF
0	62	Tangential	0.8726	0.8193
0.1	62	Tangential	0.8660	0.8293
0.2	62	Tangential	0.8490	0.7863
0.3	62	Tangential	0.8438	0.7760
0.4	62	Tangential	0.8453	0.7801
0.5	62	Tangential	0.8333	0.7410
0.6	62	Tangential	0.8166	0.7061
0.7	62	Tangential	0.7860	0.6409
0.8	62	Tangential	0.7495	0.5461
0.9	62	Tangential	0.7436	0.4959
1.0	62	Tangential	0.6917	0.3023
0.1	62	Radial	0.8683	0.8220
0.2	62	Radial	0.8645	0.8212
0.3	62	Radial	0.8644	0.8223

Table 6. Cont.

Field	Spatial Frequency (cycles/mm)	Azimuth	Design MTF	Design Plus Tolerance MTF
0.4	62	Radial	0.8661	0.8205
0.5	62	Radial	0.8545	0.7996
0.6	62	Radial	0.8458	0.7804
0.7	62	Radial	0.8337	0.7616
0.8	62	Radial	0.8156	0.7341
0.9	62	Radial	0.7745	0.6793
1.0	62	Radial	0.7228	0.6173

4. Volume and Weight of 50-Degree FOV Collimating Lens

The FOV of the 50-degree collimating lens design features an aperture stop located on the first surface of the lens, with the entrance pupil diameter of 4 mm. Four plastic aspherical lenses are used to achieve better lens quality. Shortening the gaps between the plastic lenses and the center thickness of the plastic lenses results in a smaller lens volume, as shown in Figure 6 and Table 3. The collimating optical system is divided into two parts: the lens part and the image part. The lens part consists of the first four plastic lenses at the front of the collimating optical system; the image part includes the microdisplay protection glass (BSC1-HOYA glass, 0.7 mm thick) and the microdisplay emitting surface, which is the image surface. Regarding the lens, since the aperture stop is located on the first lens, the first lens has the smallest clear aperture. The gap between the second and first lenses is tiny, which can limit the clear aperture of the second lens. The second lens is a convex lens, which effectively reduces the clear aperture of the third lens. To reduce the clear aperture of the fourth lens, we increase the image space chief ray angle θ_{CA} to $\theta_{CA} = 30.786°$ ($\theta_{CA} > 25°$) and reduce the exit pupil diameter D_{ex} to 3.131 mm, as shown in Table 7.

Table 7. Parameters for calculating exit pupil position and size.

δ	**−2.408 mm**	h′	**4.050 mm**
δ'	−7.401 mm	EFL	8.680 mm
S	2.408 mm	BFL	1.279 mm
S′	1.855 mm	θ_{CA}	30.786°
$S + \delta$	0 mm	M_T	0.7828
$S' + \delta'$	−5.516 mm	D_{ex}	3.131 mm

The FOV of the 50-degree collimating lens design has a maximum clear aperture of 6.798 mm on the second surface of the second lens. To facilitate lens fixation and clamping, the lens diameter after manufacturing is generally called the finished aperture, which is larger than the clear aperture. In accordance with the requirements of the lens manufacturer, the finished aperture is 1.2 times the clear aperture, so the maximum finished aperture of the lens is 8.156 mm. To facilitate lens assembly, the finished aperture of all four lenses is 8.156 mm. The assembly structure of the optical and mechanical components inside the lens is shown in Figure 13. Figure 13a is a perspective view of the assembled lens, the optical components are arranged from left to right as lens 1, spacer 1, lens 2, spacer 2, lens 3, spacer 3, lens 4 and retina. Figure 13b shows that the internal components of the lens include four lenses, three spacers, one retina, and barrel.

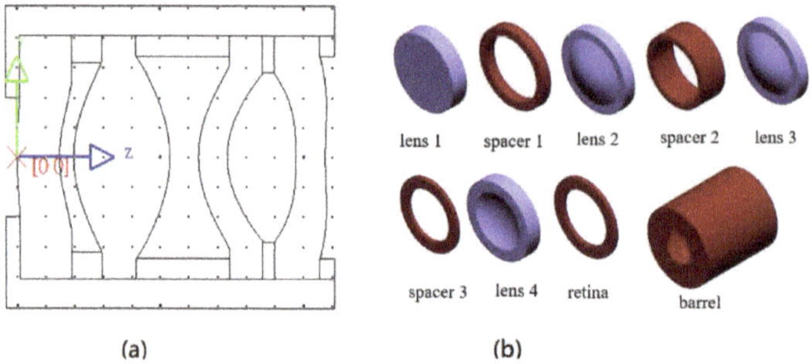

Figure 13. Assembly structure diagram of the internal optical and mechanical components of the lens. (**a**) a perspective view of the assembled lens; (**b**) internal components of the lens include four lenses, three spacers, one retina, and barrel.

Table 8 shows the volume and weight of the various optical and mechanical components within the lens. The barrel, made of PMMA plastic, serves as the lens housing. It has a mass of 0.4132 g, an outer diameter of 10.156 mm, and a length of 11.48 mm. The inner ring is hollow with an inner diameter of 8.156 mm. The barrel ring's thickness is fixed at 1 mm. The three-piece spacer structure is a mechanical component in the form of a ring that fixes the two lenses. It is made of an aluminum magnesium alloy. The retina is the ring structure that locks the final lens in place. It is made of PMMA plastic. The four lenses, three spacers, and one retina are finally assembled into the barrel, resulting in a total mass of 1.089 g. Figure 14 shows the 3D image and dimensions of the lens. The lens is a cylinder with a radius of 5.08 mm and a length of 11.48 mm; the resulting lens volume is 0.931 cm^3.

Table 8. Volume and weight of the internal optical and mechanical components of the lens.

Element Name	Material	Specific Gravity (g/cm^3)	Volume (cm^3)	Weight (g)
lens 1	OKP-1	1.22	0.1008	0.1230
spacer 1	Aluminum magnesium alloy	1.81	0.0210	0.0568
lens 2	PMMA	1.18	0.1053	0.1242
spacer 1	Aluminum magnesium alloy	1.81	0.0522	0.1410
lens 3	OKP-1	1.22	0.0907	0.1106
spacer 1	Aluminum magnesium alloy	1.81	0.0113	0.0305
lens 4	OKP-1	1.22	0.0659	0.0805
retina	PMMA	1.18	0.0079	0.0094
barrel	PMMA	1.18	0.03502	0.4132
Total				1.0892

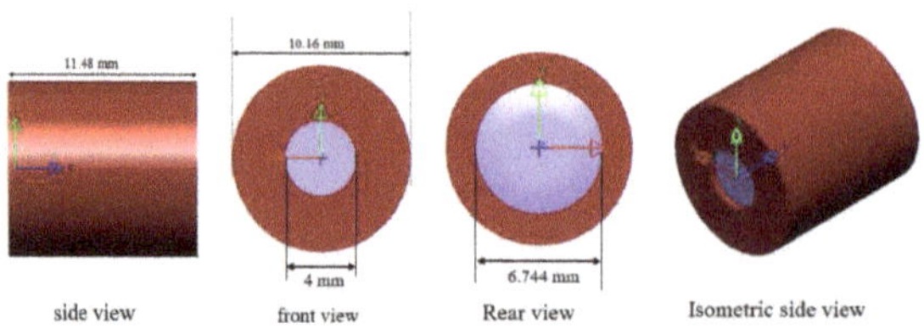

Figure 14. 3D diagram and dimensions of the lens.

As mentioned above, the 50-degree FOV collimator is designed with four aspherical lenses to balance optical quality and lens length. To reduce the size of the lens, we take the following steps:

Step 1: Select a smaller microdisplay (0.32 in).

Step 2: Set a smaller entrance pupil diameter (D_{en} = 4 mm) under a fixed F-number (F-number = 2.17) and smaller lens volume.

Step 3: Use the minimum number of lenses (4 pieces).

Step 4: Ensure that the thickness of the lenses and the gap between them is not too large.

Step 5: Place the aperture stop on the first surface of the lens to obtain the minimum clear aperture of the first lens.

Step 6: Arrange the focal length of four plastic lens symbols in the combination −, +, −, −, to reduce the clear aperture of the third lens.

Step 7: Set a smaller exit pupil diameter (D_{EX} = 3.131 mm).

Step 8: The image space chief ray angle (θ_{CA}) is larger (θ_{CA} = 30.786°), which reduces the clear aperture of the fourth lens.

The data in brackets above relates to the FOV of the 50° collimating lens design in this article.

Table 9 compares the performance of collimating lens systems in related papers on AR glasses. This paper achieves several research results: minimum optical distortion of <0.21%; the best MTF performance of >0.23 at 250 cycles/mm; the lightest lens mass of 1.09 g; and the smallest volume of 0.93 cm^3.

Table 9. Comparison of collimating lens system performance in related AR glasses papers.

Author /Year	FOV (Degree)	EPD (mm)	F-Number	Lens Number	Microdisplay Size/Resolution	MTF	Distortion (%)	Weight (g)	Volume (cm^3)
Hua [7]/2003	52.4	12	2.92	4	1.35 in 640 × 480	>0.2 at 30 cycles/mm	< 2.5%	8	3.53
Hua [8]/2007	56	10	3.2	4	1.3 in 640 × 480	>0.3 at 20 cycles/mm	<3.8%	6	3.15
Zhang [9]/2008	55	10	2.16	5	0.88 in 1280 × 1024	>0.4 at 37 cycles/mm	<4.0%	8.2	2.66
Pan [10]/2015	30	8	3.66	4	0.61 in 800 × 600	>0.4 at 30 cycles/mm	<2.7%	NR	12.45
Tsai [11]/2017	60	6	2.57	8	0.7 in 1920 × 1080	>0.29 at 62 cycles/mm	<5.0%	NR	9.76
Sun [14]/2024	30	14	1.17	9	0.35 in 1280 × 720	>0.44 at 35 cycles/mm	<2.0%	NR	25.02
Sun [15]/2025	65	10	2.05	7	1.03 in 2560 × 2560	>0.50 at 60 cycles/mm	<0.82%	NR	22.61
This work	50	4	2.17	4	0.32 in 800 × 600	>0.23 at 250 cycles/mm	<0.21%	1.08	0.93

5. Conclusions

This paper presents a compact, wide field-of-view (50°) collimating lens system designed specifically for lightguide-based augmented reality (AR) glasses. The main purpose of collimating lens design is lightweight, and the lightweight of the lens is related to the size of the microdisplay, the effective aperture of the lens, the position and size of the entrance pupil, the position and size of the exit pupil, the number of lenses, the thickness and gap between the lenses, and the arrangement of the lens symbols. First, we selected 0.32-inch microdisplays, which yielded a minimum image height of 4.05 mm. The effective aperture of the lens is related to five reference rays. Five reference rays were defined, and then the half FOV (25°) was evenly divided into 11 fields of view. Five reference rays were traced for each field of view, for a total of 55 reference rays. The effective aperture of each lens surface (including the stop and image plane) was calculated. Rays were traced along each lens surface using these 55 rays. The effective aperture of each surface was the highest ray height among these 55 rays. The optimal pupil diameter for the human eye is 4 mm, so we chose a collimator with a 4 mm entrance pupil for our design. Since the entrance pupil is located on the first lens surface, the minimum effective aperture of the first lens element is 4 mm. The exit pupil position determines the image-side chief ray angle. The effective aperture of the last lens surface is inversely proportional to the image-side chief ray angle and directly proportional to the exit pupil diameter and the image plane height. Finally, the design used four plastic aspherical lenses with a maximum lens thickness or inter-lens gap of 3.368 mm. The focal length of the four plastic lens symbols (−, +, −, −) was arranged to reduce the clear aperture of the third lens. The resulting lens weighed only 1.09 g and had a volume of 0.93 cm^3. Compared to a 2.5 g US penny, the weight of the proposed lens is only 43.6% of that, validating the feasibility of this design approach. Furthermore, compared to existing journal articles on collimating lens design for AR glasses, this lens design also achieved the lowest weight and volume.

This work builds upon previous research findings and proposes a compact design for collimating lenses for AR glasses, thereby reducing their size and weight for wearable applications. Consequently, we present a novel optical design for a compact collimating lens system designed for lightguide-based augmented reality (AR) glasses, achieving a wide field of view (FOV) of 50° using a minimal four-element aspheric plastic lens configuration. This design meets the stringent requirements of emerging wearable displays, striking a balance between high imaging performance, compactness, lightweight structure, and manufacturability.

Author Contributions: Conceptualization, W.-S.S. and Y.-L.S.; methodology, W.-S.S. and C.-L.T.; software, Y.-L.S. and Y.-S.H.; validation, W.-S.S., Y.-L.S. and C.-L.T.; formal analysis, W.-S.S. and C.-L.T.; investigation, Y.-L.S. and C.-L.T.; data curation, C.-L.T. and N.-J.C.; writing—original draft preparation, Y.-L.S.; writing—review and editing, W.-S.S. and C.-L.T.; project administration, C.-C.S.; funding acquisition, W.-S.S. All authors have read and agreed to the published version of the manuscript.

Funding: This study was supported in part by the Ministry of Science and Technology, under project numbers NSTC 113-2218-E-008-011-MBK and NSTC 113-2221-E-008-048.

Data Availability Statement: The original contributions presented in this study are included in the article. Further inquiries can be directed to the corresponding author.

Conflicts of Interest: The authors have no conflicts of interest to declare that are relevant to the content of this article.

References

1. Piao, J.A.; Li, G.; Piao, M.L.; Kim, N. Full color holographic optical element fabrication for waveguide-type head mounted display using photopolymer. *Opt. Soc. Korea* **2013**, *17*, 242–248. [CrossRef]
2. Piao, M.L.; Kim, N. Achieving high levels of color uniformity and optical efficiency for a wedge-shaped waveguide head-mounted display using a photopolymer. *Appl. Opt.* **2014**, *53*, 2180–2186. [CrossRef] [PubMed]
3. Takeyama, T. Observation Optical System. U.S. Patent 6,710,902, 23 March 2004.
4. Wu, Y.; Chen, C.P.; Zhou, L.; Li, Y.; Yu, B.; Jin, H. Design of see-through near-eye display for presbyopia. *Opt. Express* **2017**, *25*, 8937–8949. [CrossRef] [PubMed]
5. Tsai, W.P.; Tseng, Y.K.; Lee, T.X.; Sun, C.C. Volume holographic waveguide combiner design for AR glasses. *Proc. SPIE* **2024**, *12913*, 1291303.
6. Lee, T.X.; Tseng, Y.K.; Tsai, W.P.; Lin, W.K.; Zhou, S.K.; Sun, C.; Liang, Y.Y.; Yu, Y.W.; Su, W.C.; Lin, S.H.; et al. Advancing pure ray tracing for the simulation of volume holographic optical elements: Innovations in diffractive waveguide-based augmented reality systems. *Opt. Express* **2024**, *32*, 45391–45405. [CrossRef]

7. Hua, H.; Ha, Y.; Rolland, J.P. Design of an ultralight and compact projection lens. *Appl. Opt.* **2003**, *42*, 97–107. [CrossRef] [PubMed]
8. Hua, H.; Gao, C. Design of a bright polarized head-mounted projection display. *Appl. Opt.* **2007**, *46*, 2600–2610. [CrossRef] [PubMed]
9. Zhang, R.; Hua, H. Design of a polarized head-mounted projection display using ferroelectric liquid-crystal-on-silicon microdisplays. *Appl. Opt.* **2008**, *47*, 2888–2896. [CrossRef] [PubMed]
10. Pan, J.W.; Hung, H.C. Optical design of a compact see-through head mounted display with light guide plate. *J. Disp. Technol.* **2015**, *11*, 223–228. [CrossRef]
11. Tsai, C.M.; Li, J.Y.; Han, P.; Yen, C.T. Design and Evaluation of optical see-through head-mounted display with wide FOV based on dihedral corner reflector array. *IEEE Access* **2021**, *9*, 118977–118984. [CrossRef]
12. Wu, Y.; Pan, C.; Gao, Y.; Lu, C.; Zhang, Y.; Huang, Z. Design of ultra-compact augmented reality display based on grating waveguide with curved variable-period grating. *Opt. Commun.* **2023**, *529*, 128980. [CrossRef]
13. Weng, J.; Pei, C.; Yang, H.; Li, H.; Wu, R.; Liu, X. Wide-field-of-view and high-resolution waveguide display based on the coupling-collimation system. *Appl. Optics* **2024**, *63*, 4543–4551. [CrossRef]
14. Sun, W.S.; Hsu, Y.S.; Tien, C.L.; Lin, W.K.; Su, Y.L.; Yu, J.Y.; Zhou, S.K.; Liang, Y.Y.; Tsai, W.P.; Sun, C.; et al. Design and Manufacture of 30-degree Projection lens for Augmented Reality Waveguide. *Micromachines* **2024**, *15*, 1198. [CrossRef] [PubMed]
15. Sun, W.S.; Hsu, Y.S.; Su, Y.L.; Huang, G.W.; Lin, W.K.; Su, W.C.; Lin, S.H.; Lee, T.X.; Yu, Y.W.; Sun, C.C. Design of a 65-degree Collimating lens for Lightguide-based AR Glasses. *Sci. Rep.* **2025**, *15*, 24861. [CrossRef] [PubMed]
16. Mahajan, V.N. *Fundamentals of Geometrical Optics*; Chapter 6; SPIE: Bellingham, WA, USA, 2014.
17. Hecht, E. *Optics*, Chapter 6, 4th ed.; Addison-Wesley: Boston, MA, USA, 2001.
18. Synopsys Inc. Code V Document Library, Version 2023.03; Tolerance Reference Manual. Available online: https://www.synopsys.com/support/licensing-installation-computeplatforms/synopsys-documentation.html (accessed on 26 October 2025).

micromachines

Article

A Three-Terminal Si-Ge Avalanche Photodiode with a Breakdown Voltage of 6.8 V and a Gain Bandwidth Product of 1377 GHz

Chao Cheng [1], Jintao Xue [1], Xishan Yu [1], Jifang Mu [2,*] and Binhao Wang [1,3]

1 State Key Laboratory of Ultrafast Optical Science and Technology, Xi'an Institute of Optics and Precision Mechanics, Chinese Academy of Sciences, Xi'an 710119, China; chengchao2022@opt.ac.cn (C.C.); xuejintao@opt.ac.cn (J.X.); yuxishan24@mails.ucas.ac.cn (X.Y.); wangbinhao@opt.ac.cn (B.W.)
2 Photoelectric Tracking and Measurement Technology Laboratory, Xi'an Institute of Optics and Precision Mechanics, Chinese Academy of Sciences, Xi'an 710119, China
3 School of Future Technology, University of Chinese Academy of Sciences, Beijing 100049, China
* Correspondence: mujifang@opt.ac.cn

Abstract: Silicon–germanium (Si-Ge) avalanche photodiodes (APDs), fully compatible with complementary metal–oxide–semiconductor (CMOS) processes, are critical devices for high-speed optical communication. In this work, we propose a three-terminal Si-Ge APD on a silicon-on-insulator (SOI) substrate based on device simulation studies. The proposed APD employs a separate absorption and multiplication structure, achieving an ultra-low breakdown voltage of 6.8 V. The device operates in the O-band, with optical signals laterally coupled into the Ge absorption layer via a silicon nitride (Si_3N_4) waveguide. At a bias of 2 V, the APD exhibits a responsivity of 0.85 A/W; under a bias of 6.6 V, it achieves a 3-dB optoelectronic (OE) bandwidth of 51 GHz, a direct current gain of 27, and a maximum gain–bandwidth product (GBP) of 1377 GHz. High-speed performance is further confirmed through eye-diagram simulations at 100 Gbps non-return-to-zero (NRZ) and 200 Gbps four-level pulse amplitude modulation (PAM4). These results clearly show the strong potential of the proposed APD for optical communication and interconnect applications under stringent power and supply voltage constraints.

Keywords: avalanche photodiode; silicon nitride; silicon-on-insulator

1. Introduction

With the rapid advancement of artificial intelligence (AI), machine learning, and cloud computing, the demand for computing power and high-speed data transmission in large-scale data centers and high-performance computing (HPC) systems has grown exponentially [1]. However, traditional electrical interconnects suffer from high latency and excessive power consumption in long-distance transmission, making them increasingly inadequate for modern applications. In contrast, optical interconnects offer significant advantages, with latency and power consumption largely independent of transmission distance [2,3]. To address these challenges, various photonic integration platforms have been developed across different material systems, including group-III–V compounds such as indium phosphide (InP) [4] and gallium arsenide (GaAs) [5], group-IV silicon photonics (SiPh) [6–8] and planar lightwave circuits (PLCs) [9], as well as lithium niobate ($LiNbO_3$) [10]. Among them, silicon photonics (SiPh), fully compatible with standard CMOS processes, enables low-cost and high-density integration and is widely recognized

https://doi.org/10.3390/mi16111222

as a key enabler for next-generation communication systems and data transmission infrastructure [11].

As a key component in optical receiver systems, the optical receiver typically comprises fundamental modules such as couplers, demultiplexers, and APDs [12]. The performance of the APD directly determines the overall sensitivity and signal quality of the receiver, making the development of high-performance APDs fully compatible with the SiPh platform a critical research focus. Compared with Si-based demultiplexers, Si_3N_4 waveguides exhibit excellent polarization-insensitive characteristics due to their highly symmetric structure and moderate refractive index contrast with the silicon dioxide (SiO_2) cladding. In recent years, three-dimensional integrated multilayer Si_3N_4 waveguides based on the SiPh platform have been extensively investigated and experimentally demonstrated [13–15]. Edge coupler fabricated on Si_3N_4 platform exhibits excellent performance, with insertion losses below 0.5 dB per facet. However, when such demultiplexers are integrated with APDs coupled to Si waveguides, an additional transition structure is typically required to transfer the optical signal from the Si_3N_4 waveguide to the Si waveguide before detection [16]. This extra coupling step inevitably introduces additional insertion loss and polarization dependence, thereby limiting the overall system performance and integration density.

APDs with separate absorption and multiplication regions can take advantage of the high absorption efficiency of Ge in the O- and C-bands and the low impact ionization coefficient ratio of Si [17–19]. However, such devices typically require relatively high operating voltages, and most reported Si-Ge APDs operate above 10 V [20–23]. Considering that the maximum supply voltage in modern computer architectures is 12 V and that a safety margin is required in practical applications, a high APD breakdown voltage limits its applicability and increases the risk of breakdown; therefore, the breakdown voltage is preferably kept below 10 V. Fortunately, with continued in-depth research, it has been found that Si-Ge APDs with a narrow multiplication region exhibit a low temperature sensitivity of breakdown voltage. This characteristic offers a promising prospect for developing thermally robust and reliable APD devices for optical interconnects in future data centers [24,25]. Motivated by this potential, increasing attention has been devoted to reducing the breakdown voltage of such devices. To address this issue, previous studies have directly incorporated the avalanche multiplication process into the Ge region, successfully reducing the operating voltage below 10 V. Although this approach significantly lowers the bias, it often leads to increased noise, and its practical feasibility is still under investigation. More recently, three-terminal electrode structures have been proposed to further reduce the operating voltage, enabling APD operation at around 6 V and offering a more feasible solution. However, the device employs an interdigitated "finger" structure with N- and P-type doping. This doping configuration increases the PN junction area, adversely affecting the device's electrical parasitics and limiting its bandwidth, thereby constraining its potential for high-speed optical communication applications [26].

In this work, we present, for the first time, a three-terminal Si-Ge APD with lateral coupling via a Si_3N_4 waveguide. The optical signal is laterally coupled into the Ge absorption region from both sides through the Si_3N_4 waveguide, resulting in a more uniform optical field distribution within the Ge layer. Furthermore, the Si_3N_4-based lateral coupling scheme enables efficient integration with Si_3N_4-based wavelength-division multiplexers (WDMs). The choice of Si_3N_4 platform for the demultiplexer is motivated by its superior capability to achieve polarization-insensitive operation and lower insertion loss compared with demultiplexers fabricated on conventional silicon platforms [27]. In the electric field design, we introduce an innovative P+ type doping scheme in the Si waveguide beneath the Ge absorption layer, effectively shortening the carrier transit path and significantly enhancing the device's 3-dB OE bandwidth. By independently controlling the electric fields in the Ge absorption region and the multiplication region, the avalanche breakdown voltage of the device is significantly reduced to 6.8 V. As a result of these design optimizations, the APD achieves a 3-dB OE bandwidth of 51 GHz and a multiplication gain of 27 under a reverse bias voltage of 6.6 V, while exhibiting a responsivity of 0.85 A/W at 2 V. To further validate the high-speed transmission performance, eye diagram simulations were conducted at 100 Gbps using NRZ modulation and at 200 Gbps using PAM4, both of which show clearly open eyes. These results indicate that the proposed APD holds considerable potential for low-bias voltage optical communication system applications.

2. APD Design

Figure 1 illustrates the three-dimensional (3D) schematic of the proposed APD. A lateral coupling scheme via a Si_3N_4 waveguide is employed to efficiently couple the evanescent field into the Ge absorption region. The incident direction of the optical field is indicated by the red arrow. Compared to Si waveguides, Si_3N_4 waveguides offer four notable advantages [28,29]: (1) they can tolerate higher input optical power, thereby reducing the risk of waveguide damage under high-power input; (2) they exhibit lower propagation loss and smaller polarization-dependent loss, which helps mitigate the polarization sensitivity of the device; (3) they readily enable lateral coupling, allowing light to be coupled simultaneously from both sides into the Ge absorption region, thereby achieving a more uniform optical field distribution; and (4) this design enables efficient integration with polarization-insensitive Si_3N_4-based WDMs without requiring additional converters, such as vias from Si to Si_3N_4 waveguides.

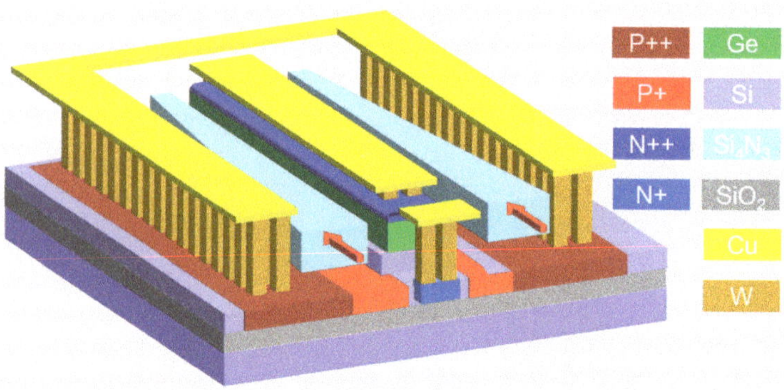

Figure 1. Three-dimensional schematic of Si-Ge APD.

The operating principle of the APD is illustrated in Figure 2. The optical signal is guided through the Si_3N_4 waveguide to a conventional 1×2 Si_3N_4 multi-mode interference (MMI), where it is split and subsequently absorbed by the Ge layer to generate photocarriers. Optical simulations were performed using three-dimensional finite-difference time-domain (3D-FDTD) methods, and the device parameters were designed with reference to the standard 130 nm SOI SiPh platform. After optimizing device performance, the determined optical structure parameters are W_1 = 0.3 μm, W_2 = 0.68 μm, L_1 = 33 μm, Gap = 0.15 μm, L_2 = 30 μm, and W_3 = 1.5 μm.

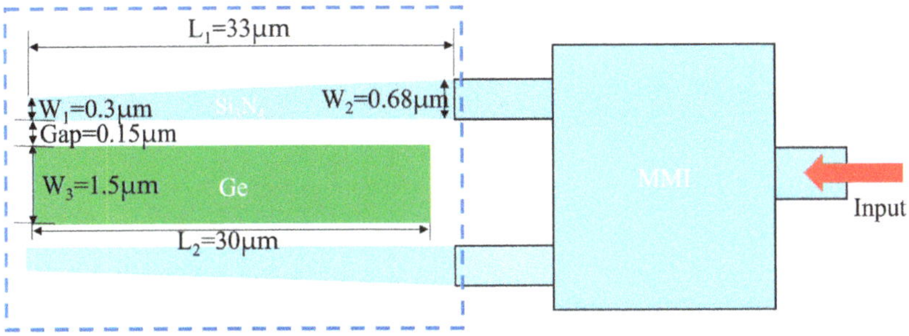

Figure 2. Simplified structural diagram of Si-Ge APD.

After completing the optical field design of the APD, the doping profile was further optimized. The overall 3D device structure is shown in Figure 1. A P^+ doping layer is introduced in the Si waveguide beneath the Ge absorption layer, with an additional third electrode connected to the P^+ region to control the avalanche electric field. The cross-sectional view of the device doping profile is illustrated in Figure 3a, where heavy doping is applied at metal-contact regions to reduce resistance. The doping concentrations are N++ (1.0×10^{20} cm^{-3}), N+ (1.0×10^{19} cm^{-3}), P++ (7.1×10^{19} cm^{-3}), and P+ (5.3×10^{18} cm^{-3}), respectively.

Figure 3b illustrates the simplified schematic of the three-terminal APD, where the three terminals—P-Ge, N-Si, and P-Si—enable independent control of two internal voltages. The voltage V1 applied between P-Ge and N-Si controls the electric field strength in the Ge region, ensuring that photogenerated carriers drift at their saturation velocity while preventing avalanche multiplication in Ge. The voltage V2 applied between N-Si and P-Si regulates the electric field strength in the Si region, thereby confining the avalanche region within the Si waveguide. Figure 3c shows the electric field distribution within the red dashed box at a 6.6 V reverse bias, with the avalanche region confined in the Si waveguide. Figure 3d shows the electric field along the orange dashed line in the P-Ge region at different reverse biases. To ensure saturated carrier drift, the electric field must exceed 1×10^6 V/m, while remaining below 1×10^7 V/m to prevent ionization in Ge [30]. Accordingly, the reverse bias voltage of the P-Ge electrode is set to 1 V, while the P-Si electrode is set to 0 V. Figure 3e depicts the electric field along the black dashed line in N-Si at a 6.6 V bias voltage, showing a field greater than 3×10^7 V/m (avalanche field threshold of silicon) and effectively confining the avalanche region within the Si layer [31].

Through optical structure optimization, the device enables efficient integration with Si_3N_4–based WDM components. Concurrently, electric field optimization allows independent control, enabling avalanche multiplication in Si while achieving optical absorption in Ge. This design fully exploits the high absorption of Ge in the O-band and leverages the low impact ionization coefficient ratio of Si to effectively suppress device noise. Furthermore, by reducing the carrier transit distance, the APD OE bandwidth is further enhanced. Taken together, these optimizations result in a device exhibiting excellent optoelectronic performance.

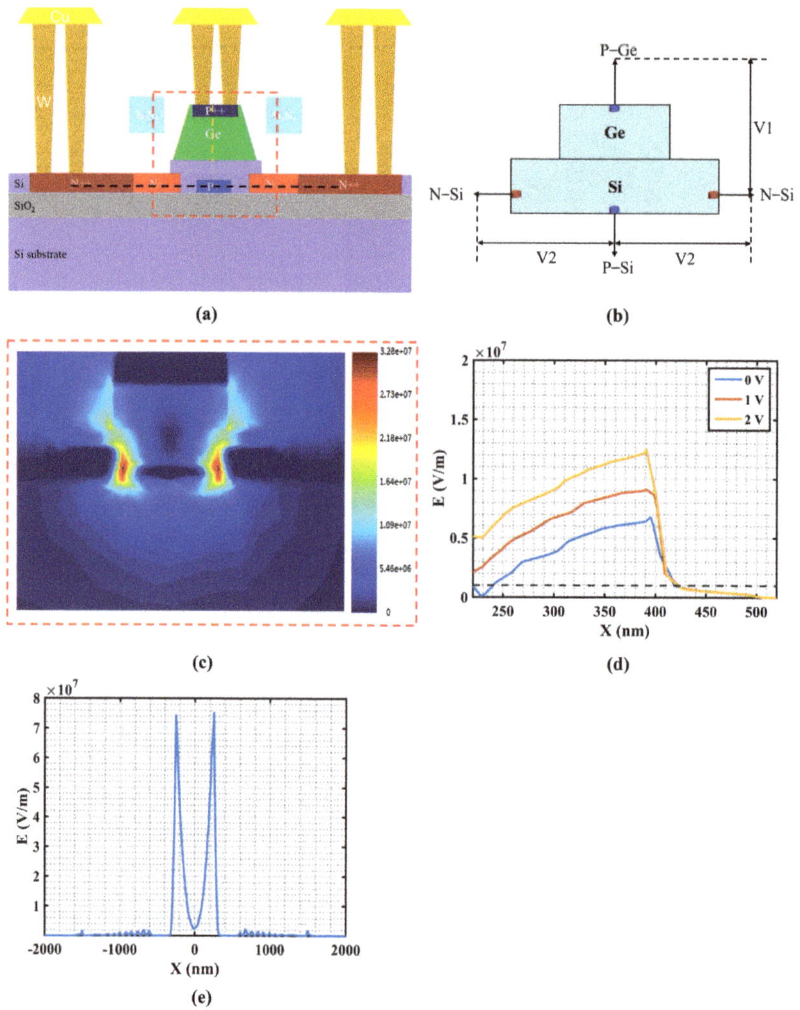

Figure 3. (**a**) Cross-sectional doping profile of the Si-Ge APD. (**b**) The simplified schematic of the three-terminal APD. (**c**) The electric field distribution in the region enclosed by the red dashed box. (**d**) Electric field along the orange dashed line in the P-Ge region under different reverse bias voltages. (**e**) Electric field along the black dashed line in the Si region at a bias voltage of 6.6 V.

3. Simulation Results

First, the optical field distribution of the APD at a wavelength of 1310 nm was simulated using FDTD. Figure 4a shows the top view of the optical field distribution, with the Ge region indicated by the red dashed box. In this APD coupling design, light is coupled from both sides, producing a uniform electric field distribution in the Ge region and thereby enhancing photodetection efficiency. Figure 4b,c present the cross-sectional optical field distributions at the input and output positions, respectively. The results indicate that the majority of the light in the Si_3N_4 waveguide is absorbed by Ge, with only a small fraction leaking outside the Ge region.

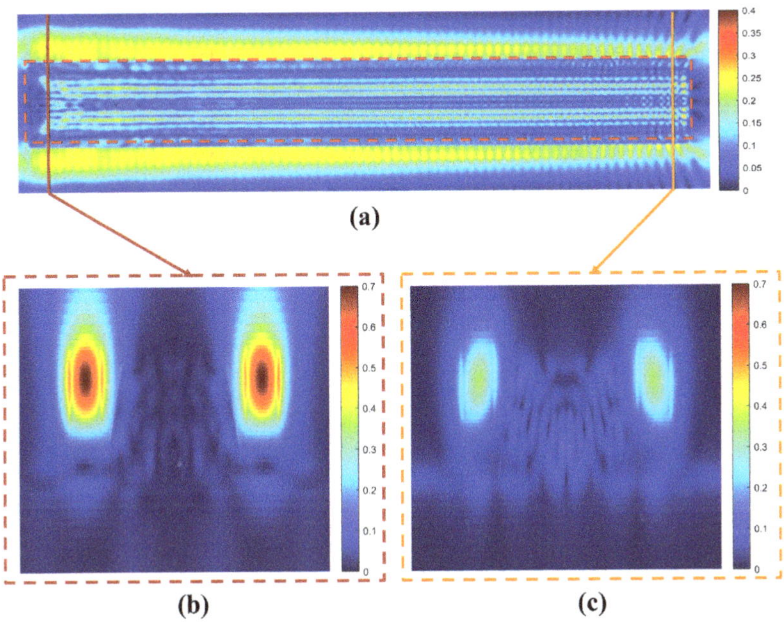

Figure 4. (**a**) Top view of the Si-Ge APD. (**b**) Cross-sectional optical field at the input of the Si_3N_4 waveguide. (**c**) Cross-sectional optical field at the output of the Si_3N_4 waveguide.

Based on the theoretical analysis described above and following the optical field analysis, the reverse bias voltage of the P–Ge electrode was set to 1V, while the P–Si electrode was set to 0 V. Under this condition, further simulations were carried out to obtain the dark current, photocurrent, gain, and small-signal response of the APD. Figure 5a shows the dark current as a function of bias voltage. The dark current of the Si-Ge APD initially increases slowly with rising reverse bias and accelerates as the breakdown voltage is approached. The device undergoes breakdown at 6.8 V, where the dark current exceeds 100 μA—significantly lower than the ~25 V breakdown voltage of conventional separate absorption charge multiplication-structured APDs [32]. Here, the breakdown voltage of Ge/Si APD is defined as the voltage at which the dark current reaches 100 μA [33]. At an operating voltage of 6.6 V, the dark current of the device is 7.31 μA. Figure 5b presents the photocurrent as a function of bias voltage under an input optical power of −21 dBm. At a reverse bias of 2 V, the device operates at unity gain, corresponding to a responsivity of 0.85 A/W. It can be observed that the photocurrent increases progressively with bias voltage, and the growth rate accelerates markedly once the avalanche threshold is exceeded. Figure 5c presents the gain curve, calculated according to Equation (1), with a maximum gain of 27 at 6.6 V reverse bias voltage. The avalanche multiplication gain $G(V)$ is defined as the difference between the photocurrent (I_p) and the dark current (I_d) at a given bias voltage, normalized to the corresponding value at the unity gain point V_0:

$$G(V) = (I_p(V) - I_d(V))/(I_p(V_0) - I_d(V_0)), \quad (1)$$

Figure 5d presents the small-signal simulation results, showing that the 3-dB OE bandwidth increases with reverse bias, reaching 51 GHz at 6.6 V. Under the same bias condition, the gain of the device is 27. Thus, the GBP, obtained as the product of gain and bandwidth, is 1377 GHz, confirming the excellent performance of the proposed APD and its suitability for high-speed data transmission.

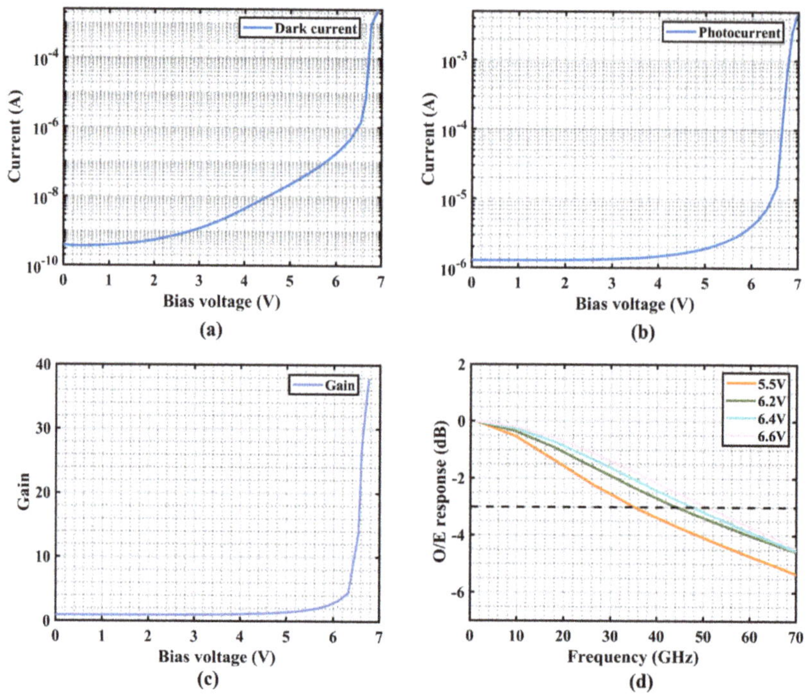

Figure 5. Performance of the Si-Ge APD. (**a**) Dark current as a function of bias voltage. (**b**) Photocurrent as a function of bias voltage. (**c**) Gain as a function of bias voltage. (**d**) Bandwidth as a function of bias voltage.

Finally, to verify the signal transmission integrity, a system-level eye diagram simulation was performed, and the results are shown in Figure 6. At a bias voltage of 6.6 V and a temperature of 25 °C, the device exhibits a dark current of 7.31 μA, a gain of 27, and a 3-dB bandwidth of 51 GHz. Under these conditions, the static power consumption of the device is 48.2 μW, while the dynamic power consumption reaches 1.2 mW at an input optical power of −21 dBm. The APD under these operating conditions was used for the system-level simulation. A 1310 nm laser source was employed as the optical input, and a high-speed radio frequency (RF) signal was applied through an intensity modulator. The modulated optical signal was then injected into the APD to generate eye diagrams. To fully demonstrate the high-speed transmission capability of the proposed APD, no additional noise source was included in the simulation. Figure 6a,c show the input optical eye diagrams for 100 Gbps NRZ and 200 Gbps PAM4 signals, respectively, while Figure 6b,d present the corresponding output eye diagrams after the APD detection, with signal-to-noise ratios (SNRs) of 12.5 dB and 3.49 dB, respectively. The results indicate that the high-speed modulated signals maintain clear eye openings after passing through the APD, demonstrating excellent signal integrity during high-speed transmission. Compared with the Si-Ge PD without avalanche gain, the internal multiplication effect of the APD contributes to an improved SNR under low-input optical power.

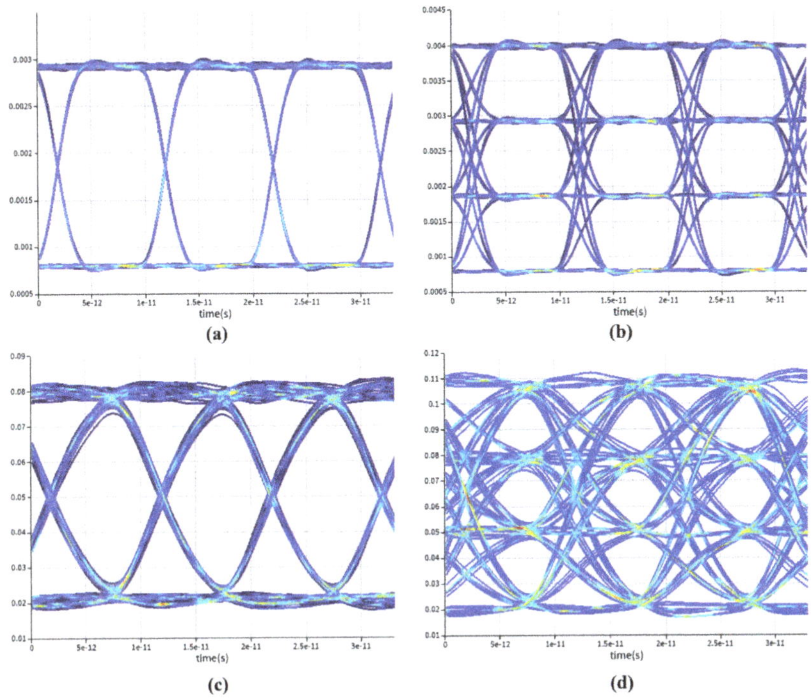

Figure 6. For the Si-Ge APD, eye diagrams for 100 Gbps NRZ signals. (**a**) Original optical signal eye diagram. (**c**) Eye diagram after detection, eye diagrams for 200 Gbps PAM4 signals. (**b**) Original optical signal eye diagram. (**d**) Eye diagram after detection.

4. Discussion

In recent years, various advanced modulation formats and multidimensional multiplexing techniques have been widely adopted to further enhance the transmission rate and capacity of optical communication systems, fully exploiting their bandwidth potential. Currently, Si_3N_4 waveguides can be efficiently integrated on SiPh platforms, enabling direct coupling of optical signals into devices. However, most existing APDs still rely on Si waveguide inputs, necessitating a converter between the Si_3N_4 and Si waveguides [16], which not only increases system loss but also exacerbates polarization dependence. Therefore, direct coupling via Si_3N_4 waveguides remains a promising avenue for further exploration.

Meanwhile, in high-speed optical interconnect systems such as data centers, the bias voltage of APDs remains a critical bottleneck. As summarized in Table 1, conventional APDs typically require operating voltages of 10 V or higher [2,17,34,35], which severely limits further integration. To address this, considerable efforts have been made to reduce the operating voltage. For instance, a three-terminal APD proposed in 2019 lowered the operating voltage to 6 V; however, due to its doping configuration, the device's capacitance and other electrical characteristics were affected, further limiting the bandwidth. Additionally, its responsivity was relatively low, with only 0.48 A/W at a reverse bias of 2 V in the C-band [26].

Compared with previous three-terminal designs, the optimizations in this work focus on two key aspects: (1) employing lateral coupling through a Si_3N_4 waveguide, which mitigates optical absorption induced by heavy doping, thereby preventing responsivity degradation while also simplifying integration with Si_3N_4 waveguide-based wavelength-division multiplexers (WDMs); and (2) introducing doping beneath the Ge absorption region to shorten the carrier collection path, which contributes to bandwidth enhancement.

Table 1. Summary of the state-of-the-art Si-Ge APDs.

Reference	Avalanche Region	λ (nm)	V_{bias} (V)	Id (μA)	Gain	BW (GHz)	GBP (GHz)	Bit Rate (Gbps)
[2]	Si	1310	−12.5	~8.4	244	29	7078	112 (PAM4)
[17]	SI	1310	−12	~100	11	27	300	50 (OOK)
[26]	Si	1550	−6	~10	15	18.9	284	35 (OOK)
[34]	Si	1310	−17	~0.12	-	56	-	-
[35]	Si	1310	−11	~3.5	8	17	136	56 (OOK)
This Work *	Si	1310	−6.6	7.31	27	51	1377	200 (PAM4)

* Simulation results.

Although Si_3N_4 waveguide integration on Si platforms is already mature and the optical structures are relatively easy to fabricate, our design introduces an innovative P^+ doping layer in the Si waveguide beneath the Ge region, along with an additional electrode. Compared with conventional two-electrode APDs [36,37], this device presents certain fabrication challenges; thus, the electrode placement still has room for optimization, and further refinement of the doping profile to simplify fabrication constitutes future work. Despite these challenges, the proposed APD remains highly attractive for achieving lower operating voltages.

5. Conclusions

In this study, we simulated the design of an O-band three-terminal Si-Ge APD, where the optical signal is laterally coupled into the Ge absorption region via a Si_3N_4 waveguide. At a reverse bias voltage of 2 V, the device exhibits a responsivity of 0.85 A/W; under 6.6 V bias, the bandwidth reaches 51 GHz with a DC gain of 27, yielding a GBP of 1377 GHz. Eye-diagram simulations for 100 Gbps NRZ and 200 Gbps PAM4 signals further confirm its high-speed performance. The Si-Ge APD is highly compatible with Si_3N_4 WDM and can achieve avalanche multiplication at relatively low bias voltage, demonstrating broad potential for high-speed optical communication applications.

Author Contributions: Conceptualization, methodology, resources, writing—review and editing and funding acquisition, B.W. and J.M.; software, validation, writing—original draft preparation and visualization, C.C., J.X. and X.Y. All authors have read and agreed to the published version of the manuscript.

Funding: This research was funded by the National Key Research and Development Program of China (2022YFB2803100).

Data Availability Statement: Data underlying the results presented in this paper are not publicly available at this time but may be obtained from the authors upon reasonable request.

Conflicts of Interest: The authors declare no conflicts of interest.

References

1. Xue, J.; Wang, B.; Chen, Y.; Chen, S.; Wu, J.; Cheng, C.; Bao, S.; Qi, N.; Zhang, W. A 4 × 112 Gb/s Ultra-Compact Polarization-Insensitive Silicon Photonics WDM Receiver with CMOS TIA for Co-Packaged Optics and Optical I/O. *J. Light. Technol.* **2024**, *12*, 6028–6035. [CrossRef]
2. Cheng, C.; Xue, J.; Bao, S.; Liu, Q.; Zhang, W.; Wang, B. A 112 Gbps Silicon-Germanium Avalanche Photodiode with Ultra-High Gain-Bandwidth Product. In Proceedings of the Optical Fiber Communications Conference and Exhibition (OFC), San Francisco, CA, USA, 30 March–3 April 2025.

3. Timurdogan, E.; Su, Z.; Shiue, R.-J.; Byrd, M.J.; Poulton, C.V.; Jabon, K.; DeRose, C.; Moss, B.R.; Hosseini, E.S.; Duzevik, I.; et al. 400g silicon photonics integrated circuit transceiver chipsets for cpo, obo, and pluggable modules. In Proceedings of the 2020 Optical Fiber Communications Conference and Exhibition (OFC), San Diego, CA, USA, 8–12 March 2020.
4. Smit, M.; Leijtens, X.; Ambrosius, H.; Bente, E.; Van Der Tol, J.; Smalbrugge, B.; De Vries, T.; Geluk, E.-J.; Bolk, J.; Van Veldhoven, R.; et al. An introduction to InP-based generic integration technology. *Semicond. Sci. Technol.* **2014**, *29*, 083001. [CrossRef]
5. Chen, G.; Goyvaerts, J.; Kumari, S.; Van Kerrebrouck, J.; Muneeb, M.; Uvin, S.; Yu, Y.; Roelkens, G. Integration of high-speed GaAs metalsemiconductor-metal photodetectors by means of transfer printing for 850 nm wavelength photonic interposers. *Opt. Express* **2018**, *26*, 6351–6359. [CrossRef]
6. Zhou, D.; Sun, C.; Lai, Y.; Yu, Y.; Zhang, X. Integrated silicon multifunctional mode-division multiplexing system. *Opt. Express* **2019**, *27*, 10798–10805. [CrossRef] [PubMed]
7. Zhou, D.; Yu, Y.; Zhang, X. Parallel radio-frequency signal-processing unit based on mode multiplexed photonic integrated circuit. *Opt. Express* **2018**, *26*, 20544–20549. [CrossRef] [PubMed]
8. Dong, P.; Xie, C.; Buhl, L.L. Monolithic polarization diversity coherent receiver based on 120-degree optical hybrids on silicon. *Opt. Express* **2014**, *22*, 2119–2125. [CrossRef] [PubMed]
9. Calafiore, G.; Koshelev, A.; Dhuey, S.; Goltsov, A.; Sasorov, P.; Babin, S.; Yankov, V.; Cabrini, S.; Peroz, C. Holographic planar lightwave circuit for on-chip spectroscopy. *Light Sci. Appl.* **2014**, *3*, e203. [CrossRef]
10. Rao, A.; Fathpour, S. Compact Lithium Niobate Electrooptic Modulators. *IEEE J. Sel. Top. Quant.* **2018**, *24*, 1–14. [CrossRef]
11. Bowers, J.E.; Liu, A.Y. A Comparison of Four Approaches to Photonic Integration. In Proceedings of the Optical Fiber Communication Conference and Exhibition (OFC), Los Angeles, CA, USA, 19–23 March 2017.
12. Xue, J.; Wu, J.; Cheng, C.; Zhang, W.; Wang, B. A 4 × 112Gbps Compact Polarization-Insensitive Silicon Photonic WDM Receiver. In Proceedings of the Optical Fiber Communications Conference and Exhibition (OFC), San Diego, CA, USA, 24–28 March 2024.
13. Caut, A.; Shekhawat, V.; Torres-Company, V.; Karlsson, M. Polarization-Insensitive Silicon Nitride Photonic Receiver at 1 μm for Optical Interconnects. *IEEE Photonics J.* **2024**, *16*, 1–7.
14. Mikkelsen, J.C.; Bois, A.; Lordello, T.; Mahgerefteh, D.; Menezo, S.; Poon, J.K.S. Polarization-insensitive silicon nitride mach-zehnder lattice wavelength demultiplexers for CWDM in the O-band. *Opt. Express* **2018**, *26*, 30076–30084. [CrossRef]
15. He, A.; Guo, X.; Wang, T.; Su, Y. Ultracompact fiber-to-chip metamaterial edge coupler. *ACS Photonics* **2021**, *8*, 3226–3233. [CrossRef]
16. Sacher, W.D.; Mikkelsen, J.C.; Dumais, P.; Jiang, J.; Goodwill, D.; Luo, X.; Huang, Y.; Yang, Y.; Bois, A.; Lo, P.G.-Q.; et al. Tri-layer silicon nitride-on-silicon photonic platform for ultra-low-loss crossings and interlayer transitions. *Opt. Express* **2017**, *25*, 30862–30875. [CrossRef]
17. Srinivasan, S.A.; Berciano, M.; De Heyn, P.; Lardenois, S.; Pantouvaki, M.; Van Campenhout, J. 27 Ghz silicon-contacted waveguide-coupled Ge/Si Avalanche Photodiode. *J. Light. Technol.* **2020**, *38*, 3044–3050. [CrossRef]
18. Wang, B.; Huang, Z.; Yuan, Y.; Liang, D.; Zeng, X.; Fiorentino, M.; Beausoleil, R.G. 64 gb/s low-voltage waveguide si-ge avalanche photodiodes with distributed bragg reflectors. *Photonics Res.* **2020**, *8*, 1118–1123. [CrossRef]
19. Cao, H.; Xiang, Y.; Sun, W.; Xie, J.; Guo, J.; Dai, D. Waveguide Ge/Si Avalanche Photodetector with Ultra-High Gain-Bandwidth Product of 1440GHz. In Proceedings of the 2023 Asia Communications and Photonics Conference/2023 International Photonics and Optoelectronics Meetings (ACP/POEM), Wuhan, China, 4–7 November 2023.
20. Xiang, Y.; Cao, H.; Liu, C.; Guo, J.; Dai, D. High speed waveguide ge/si avalanche photodiode with a gain-bandwidth product of 615 GHz. *Optica* **2022**, *9*, 762–769. [CrossRef]
21. Wei, Z.; Zhang, Q.; Huang, Q.; Fang, S.; Jiang, X.; Zhang, S.; Shi, Y.; Wang, Y.; Zhou, G.; Yang, J.; et al. Ultrahigh gain-bandwidth product Ge/Si avalanche photodetector assisted by nonuniform electric field distribution. In Proceedings of the European Conference on Optical Communications (ECOC), Frankfurt, Germany, 22–26 September 2024.
22. Kang, Y.; Liu, H.-D.; Morse, M.; Paniccia, M.J.; Zadka, M.; Litski, S.; Sarid, G.; Pauchard, A.; Kuo, Y.-H.; Chen, H.-W.; et al. Monolithic germanium/silicon avalanche photodiodes with 340 GHz gain–bandwidth product. *Nat. Photonics* **2009**, *3*, 59–63. [CrossRef]
23. Wei, Z.; Zhang, Q.; Huang, Q.; Fang, S.; Jiang, X.; Zhang, S.; Shi, Y.; Wang, Y.; Zhou, G.; Yang, J.; et al. High-gain Ge/Si avalanche photodetector enhanced by distributed structure. In Proceedings of the Optical Fiber Communications Conference and Exposition (OFC), San Francisco, CA, USA, 30 March–3 April 2025.
24. Tsiara, A.; Berciano, M.; Yudistira, D.; Loo, R.; Musibau, S.; Verheyen, P.; Ferraro, F.; Ban, Y.; Croes, K.; Van Campenhout, J. 50 Gbps vertical separate absorption charge multiplication Ge/Si avalanche waveguide photodetectors integrated in a 300-mm Si photonics platform. In Proceedings of the 49th European Conference on Optical Communications (ECOC), Glasgow, UK, 1–5 October 2023.
25. Yuan, Y.; Huang, Z.; Wang, B.; Sorin, W.V.; Zeng, X.; Liang, D.; Fiorentino, M.; Campbell, J.C.; Beausoleil, R.G. 64 Gbps PAM4 Si-Ge Waveguide Avalanche Photodiodes With Excellent Temperature Stability. *J. Light. Technol.* **2020**, *17*, 4857–4866. [CrossRef]

26. Zeng, X.; Huang, Z.; Wang, B.; Liang, D.; Fiorentino, M.; Beausoleil, R.G. Silicon–germanium avalanche photodiodes with direct control of electric field in charge multiplication region. *Optica* **2019**, *6*, 772–777. [CrossRef]
27. Fernández-Hinestrosa, A.; Luque-González, J.M.; Cheben, P.; Schmid, J.H.; Sánchez-Postigo, A.; Wangüemert-Pérez, J.G.; Molina-Fernández, I.; Ortega-Moñux, A. Polarization-independent complex bragg grating filters on silicon nitride. *Laser Photonics Rev.* **2025**, e02114. [CrossRef]
28. Cheng, C.; Xue, J.; Yu, Z.; Wu, J.; Bao, S.; Wang, B. A High-Power Lateral p-i-n Silicon-Germanium Photodiode. In Proceedings of the 2024 Photonics &Electromagnetics Research Symposium (PIERS), Chengdu, China, 21–25 April 2024.
29. Hu, X.; Wu, D.; Zhang, H.; Li, W.; Chen, D.; Wang, L.; Xiao, X.; Yu, S. High-speed and high-power germanium photodetector with a lateral silicon nitride waveguide. *Photonics Res.* **2021**, *9*, 749–756. [CrossRef]
30. Shi, Y.; Li, X.; Chen, G.; Zou, M.; Cai, H.; Yu, Y.; Zhang, X. Avalanche photodiode with ultrahigh gain–bandwidth product of 1,033 GHz. *Nat. Photon.* **2024**, *18*, 610–616. [CrossRef]
31. Sze, S.M.; Ng, K.K. *Physics of Semiconductor Devices*, 3rd ed.; John Wiley & Sons: Hoboken, NJ, USA, 2007.
32. Huang, Z.; Li, C.; Liang, D.; Yu, K.; Santori, C.; Fiorentino, M.; Sorin, W.; Palermo, S.; Beausoleil, R.G. 25 Gbps low-voltage waveguide Si–Ge avalanche photodiode. *Optica* **2016**, *3*, 793–798. [CrossRef]
33. Hu, X.; Zhang, H.; Wu, D.; Chen, D.; Wang, L.; Xiao, X.; Yu, S. High-performance germanium avalanche photodetector for 100 Gbit/s photonics receivers. *Opt. Lett.* **2021**, *46*, 3837–3840. [CrossRef]
34. Huang, M.; Cai, P.; Li, S.; Hou, G.; Zhang, N.; Su, T.-I.; Hong, C.-Y.; Pan, D. 56GHz waveguide Ge/Si avalanche photodiode. In Proceedings of the Optical Fiber Communications Conference and Exposition (OFC), San Diego, CA, USA, 11–15 March 2018.
35. Srinivasan, S.A.; Lambrecht, J.; Guermandi, D.; Lardenois, S.; Berciano, M.; Absil, P. 56 Gb/s NRZ O-Band hybrid BiCMOS-Silicon photonics receiver using Ge/si avalanche photodiode. *J. Light. Technol.* **2020**, *39*, 1409–1415. [CrossRef]
36. Zhu, Y.; Liu, Z.; Niu, C.; Pang, Y.; Zhang, D.; Liu, X.; Zheng, J.; Zuo, Y.; Xue, H.; Cheng, B. High-speed and high-power germanium photodetector based on a trapezoidal absorber. *Opt. Lett.* **2022**, *47*, 3263–3266. [CrossRef]
37. Yuan, Y.; Huang, Z.; Zeng, X.; Liang, D.; Sorin, W.V.; Fiorentino, M.; Beausoleil, R.G. High responsivity Si-Ge waveguide avalanche photodiodes enhanced by loop reflector. *IEEE J. Sel. Top. Quantum Electron.* **2022**, *28*, 1–8. [CrossRef]

micromachines

Article

Optimization of Overdriving Pulse for Luminance Stability of Electrowetting Displays

Yanjun Yang [1,2], Zichuan Yi [1,*], Wanzhen Xu [2], Jiashuai Wang [3], Qingsong Lu [1], Qifu Liu [1], Liming Liu [1] and Feng Chi [1]

1 School of Electronic Information, University of Electronic Science and Technology of China, Zhongshan Institute, Zhongshan 528402, China; 2024024416@m.scnu.edu.cn (Y.Y.); 2024024402@m.scnu.edu.cn (Q.L.); 113137360827@163.com (Q.L.); liulmxps@126.com (L.L.); chifeng@semi.ac.cn (F.C.)

2 Guangdong Provincial Key Laboratory of Optical Information Materials and Technology, South China Academy of Advanced Optoelectronics, South China Normal University, Guangzhou 510006, China; 2024010350@m.scnu.edu.cn

3 School of Integrated Circuit Science and Engineering (Exemplary School of Microelectronics), University of Electronic Science and Technology of China, Chengdu 611731, China; 202511012418@std.uestc.edu.cn

* Correspondence: yizichuan@zsc.edu.cn; Tel.: +86-137-6066-9805

Abstract: As a reflective display technology, electrowetting displays (EWDs) have the advantages of a paper-like appearance, fast response speed, and full-color capability. However, the use of an overdriving voltage to improve the response speed of EWDs can cause fluctuations in display luminance, which manifest as glitches in the luminance change curve. In order to eliminate this luminance instability phenomenon, a new driving pulse is proposed, which consists of an overdriving phase, a switching phase, and a driving phase. Firstly, a simplified equivalent circuit model is proposed to apply a target voltage in the driving phase without break down of the hydrophobic insulating layer. Secondly, a COMSOL (Version 6.3) two-dimensional model is established to simulate the oil contraction process and conduct comparisons, so as to ensure the effectiveness of the overdriving pulse. Then, the overdriving phase is applied to improve oil response speed, and a linear function is used in the switching phase to alleviate glitch phenomena. Moreover, the influences of overdriving voltage, overdriving time, and linear switching time on the luminance curve are analyzed by charge trapping theory in order to obtain optimal performance. The experimental results show that the glitch phenomenon is eliminated effectively, and the luminance of the EWD is increased by 1.02% and 1.96% compared with the step switching pulse and PWM pulse, respectively, while the response time is shortened by 1.82% and 8.05% compared with the step switching pulse and PWM pulse, respectively.

Keywords: electrowetting display (EWD); overdriving pulse; electrowetting principle; linear conversion; Luminance stability

1. Introduction

Electrowetting display (EWDs) [1], as emerging reflective display technologies, have become a research hotspot in the field of display technology due to their significant advantages, such as high reflectivity [2], fast response speed, and low power consumption [3,4]. However, although EWD technology has many potential advantages, it still faces many challenges in terms of practical application, such as luminance reduction caused by charge trapping [5]. Therefore, it is of great significance to solve these problems by optimizing the overdriving pulse in order to promote the commercialization and industrialization of EWD technology. So far, most studies on EWDs mainly aim to optimize or design a new driving

 https://doi.org/10.3390/mi16101085

pulse to reduce oil splitting [6–8], oil backflow [9–11], charge trapping [12–14], or optimize pixel structure [15–17], aiming to improve performance metrics such as aperture ratio, response speed, and luminance stability. Among them, response speed is an important indicator. To improve the response speed of droplets or pixels, researchers have proposed a variety of gradually optimized driving strategies and methods. Early research mainly focused on enhancing response performance by applying an overdriving voltage. For example, by studying the spreading behavior and response time of droplets under different DC and overdriving conditions, appropriate overdriving signals were selected to improve the response speed of viscous fluids [18]. Similarly, regarding the issue of response time for the rising use of electrowetting-driven low-viscosity liquid columns, studies have found that, compared with DC driving, the overdriving pulse can effectively reduce the response time [19]. Meanwhile, to further shorten the response time of the devices, an input voltage function with multiple exponential rise times was adjusted [20]. With the deepening of research, single-voltage control gradually evolved into multi-stage pulse design to balance speed and stability. For instance, a driving pulse based on the overdriving voltage and charge trapping theory was proposed, which shortens the response time of EWDs while taking charge trapping control into account [21]. To suppress oil backflow in the active matrix, a separated reset pulse was designed. In the first stage, an overdriving voltage is used to increase the response speed, while the voltage of all pixels is inverted to achieve a stable state [22]. Aiming at the impact of the hysteresis effect, an equivalent driving pulse design method with an overdriving voltage was proposed, which effectively improved the response performance [23]. Based on the analysis of the oil backflow phenomenon, a new driving pulse using an exponential function was proposed. This pulse increases the voltage exponentially, starting from the threshold voltage, and combines with overvoltage driving to suppress splitting and improve response speed [24]. To further break through performance bottlenecks, research has begun to focus on both pixel structure and pulse. At the pixel structure level, a multi-electrode pixel structure was proposed. This multi-electrode pixel structure divides the pixel into four square sub-electrodes, and a three-dimensional EWD simulation model was established for verification, improving the response speed through physical design [25]. At the pulse design level, to suppress charge trapping and mitigate the impact of leakage current, an efficient driving pulse was proposed. It uses an overdriving voltage in the driving stage to accelerate the response speed and maintains the target voltage in the stable stage to sustain the state [26]. Similarly, a multi-channel DC overdriving pulse was proposed, which divides the pulse into a start-up driving stage and a stable driving stage. The introduction of an overdriving voltage in the start-up stage significantly improved the response speed and enhanced grayscale consistency [27]. In addition, a new composite function pulse based on a sampling function has been proposed in recent years, which consists of an initial driving stage and a backflow suppression stage. In the initial driving stage, an overdriving voltage is used to quickly drive the pixel, and the overdriving voltage is attenuated to the target voltage, which not only improves the response speed but also effectively prevents damage to the pixel [28]. It can be seen that the above studies improve the response speed in different ways, most of which involve adding to the overdriving voltage to improve the response speed. However, when using overdriving voltage, the phenomenon of glitches in the luminance curve is ignored, which affects the stability of the luminance of the display.

To eliminate the glitch phenomenon in the luminance curve caused by overdriving voltage, a new pulse is proposed in this paper based on overdriving pulse and the electrowetting principle. Then, a COMSOL two-dimensional model is established to verify the feasibility of the designed pulse, and the relationship between overdriving voltage and overdriving time is studied. Lastly, a linear switching phase is designed to alleviate abrupt voltage changes and improve the luminance stability of the EWD.

2. Theory

2.1. Principle of EWD

The pixel structure of EWDs mainly consists of the top glass, ITO electrode, pixel wall, polar liquid, colored oil, hydrophobic insulating layer (fluoropolymer), and bottom substrate. When no voltage is applied, the oil covers the whole bottom of the pixel, the pixel presents the color of the oil at this time, and the EWD is in the "off" state, as shown in Figure 1a,c. On the contrary, when a certain voltage is applied between the electrodes, the surface tension of the liquid changes and the liquid is pushed or diffused, so that the surface of the hydrophobic insulating layer can be displayed, as shown in Figure 1b,d.

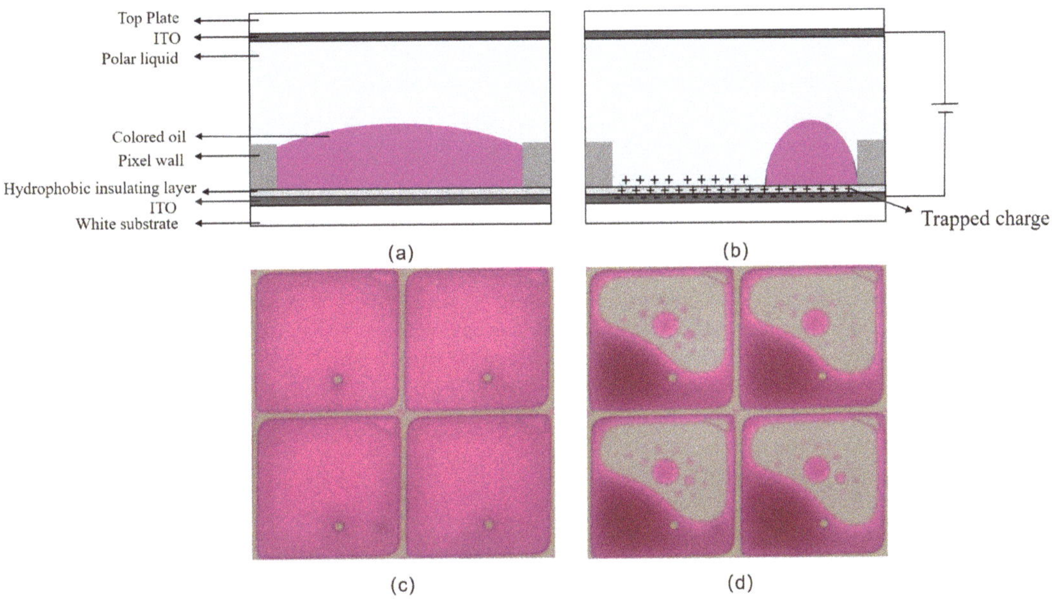

Figure 1. Structure of EWD pixels. (**a**) EWD in the "off" state; (**b**) EWD in the "on" state; (**c**) physical drawing of EWD in "off" state; (**d**) physical drawing of EWD in "on" state.

In order to analyze the relationship between driving voltage and oil movement, and given that the capacitance of the pixel wall can be neglected due to the low conductivity of the photoresist material, a simplified equivalent circuit model [29] is proposed. The current in the hydrophobic insulating layer and oil film obeys Ohm's law, so that the effective capacitance and effective current of a single pixel can be obtained, as shown in Equations (1) and (2).

$$C = C_{oil} + C_D = \varepsilon_0 (\frac{\varepsilon_{oil} S_{oil}}{h} + \frac{\varepsilon_D S_{pixel}}{d}) \tag{1}$$

$$I_C = C\frac{dV}{dt} \tag{2}$$

where C and I_C are, respectively, the effective capacitance and effective current of a single pixel; C_{oil} and C_D are, respectively, the capacitances of the oil and the hydrophobic insulating layer; ε_0 is the vacuum dielectric constant; ε_{oil} and ε_D are, respectively, the dielectric constants of the oil and the hydrophobic insulating layer, S_{oil} is the area where oil shrinks to the corners of a pixel; and S_{pixel} is the area of a single pixel. h and d are, respectively, the thicknesses of the oil and the hydrophobic insulating layer, and V is the applied driving voltage. In addition, the correspondence between contact angle and driving voltage can be described by the Lippmann–Young equation [30], which can be modified according to refs. [30,31], as shown in Equation (3).

$$\cos\theta = 1 - \frac{CV^2}{2\gamma} \tag{3}$$

where θ is the contact angle and γ is the interfacial tension between the colored oil and the conductive solution. It can be seen that the contact angle is proportional to the driving voltage. At the same time, the relationship between the thickness of the hydrophobic insulating layer and the driving voltage can be obtained from Equations (1) and (3), as shown in Equation (4).

$$V = \sqrt{\frac{2(1-\cos\theta)\gamma}{C}} \sim \sqrt{d} \tag{4}$$

It can be seen that the driving voltage is proportional to the thickness of the hydrophobic insulating layer. When the driving voltage increases to a critical value, a phenomenon occurs where the contact angle no longer decreases as the voltage continues to rise, which is known as contact angle saturation [30]. Subsequent further increases in the driving voltage will disrupt the stability of the oil, leading to the failure of the EWD. When no voltage is applied, the oil in the pixel should spread evenly across the bottom. This phenomenon, as observed with the naked eye and under a microscope, is shown in Figure 2a,c, respectively. However, after the EWD fails, colored oil droplets remain trapped in the corners of the pixel and cannot spread back or even overflow into adjacent pixels, forming scattered white spots. This phenomenon, as observed with the naked eye and under a microscope, after failure is shown in Figure 2b,d, respectively, where the regions of obvious failure are highlighted with green frames. Therefore, an excessively high driving voltage should not be applied during the experiment.

In addition, when a driving voltage is applied between the electrodes of the EWD, ions in the polar liquid are pulled toward the hydrophobic insulating layer by an electric field. These ions may enter the layer and become trapped, leading to charge accumulation. At this time, a reverse electric field is generated, as shown in Equation (5) and Figure 1b.

$$V_T = \frac{Q_T}{C_D} \tag{5}$$

where V_T is the electric potential generated by charge trapping and Q_T is the amount of trapped charge. Equation (3) is modified to include the effect of charge trapping [31], as shown in Equation (6).

$$\cos\theta = 1 - \frac{C(V - V_T)^2}{2\gamma} \tag{6}$$

It can be seen that as the drive voltage increases the influence of charge trapping becomes greater.

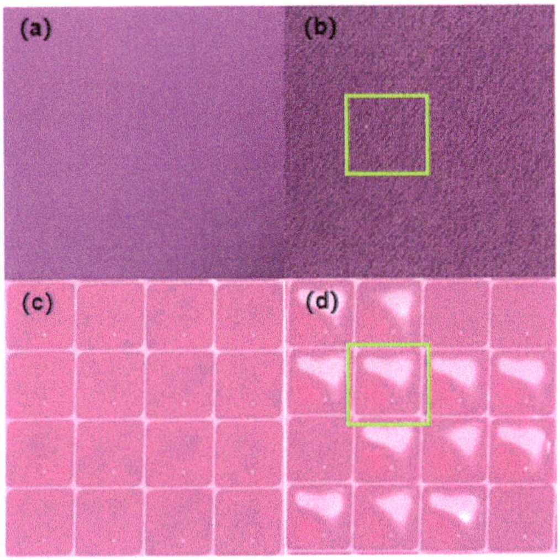

Figure 2. (**a**) An undamaged EWD under visual inspection; (**b**) a damaged EWD under visual inspection; (**c**) the pixel structure of the undamaged EWD under a microscope; (**d**) the pixels of the damaged EWD under a microscope.

2.2. *EWD Simulation Model*

2.2.1. Governing Equation

To investigate the dynamic response characteristics of the oil interface and the generation mechanism of the glitch phenomenon under the action of the overdriving voltage from the perspective of fluid dynamics, the phase-field method was used to study the dynamic process of the two-phase flow interface [32–35]. The governing equation of this model is shown in Equation (7).

$$\begin{aligned} &\tfrac{\partial \varphi}{\partial t} + u \cdot \nabla \varphi = \nabla \cdot \tfrac{\gamma \lambda}{\varepsilon_{pf}^2} \nabla \psi \\ &\psi = -\nabla \cdot \varepsilon_{pf}^2 \nabla \varphi + (\varphi^2 - 1)\varphi + \tfrac{\varepsilon_{pf}^2}{\lambda} \cdot \tfrac{\partial f}{\partial \varphi} \\ &\lambda = \tfrac{3\varepsilon_{pf}}{\sqrt{8}} \end{aligned} \tag{7}$$

where u is the velocity of the fluid, γ is the mobility parameter, ∇ is the gradient operator, ε_{pf} is the capillary width, and λ is the energy density; both φ and ψ are phase-field variables, where φ is used to distinguish and track two immiscible fluids, with $\varphi = 1$ representing oil and $\varphi = -1$ representing water. In this model, the motion of the liquid on the solid surface is indirectly tracked by solving the variables ψ and φ in the equation [36,37]. Also, to describe the dynamic motion of the immiscible two-phase flow, the transfer of mass and momentum is governed by the incompressible Navier–Stokes equations [38], as shown in Equations (8) and (9).

$$\rho\left(\frac{\partial u}{\partial t} + u \cdot \nabla u\right) = -\nabla p + \nabla \cdot \left(\mu\left(\nabla u + (\nabla u)^T\right) - \frac{2}{3}u(\nabla \cdot u)I\right) + F \tag{8}$$

$$F = F_{st} + \rho g + F_{vf} \tag{9}$$

where p represents the pressure of the fluid, ρ represents the density of the fluid, μ represents the dynamic viscosity of the fluid, F represents the external force, I is the unit matrix, and each term in Equation (8) represents inertial force, pressure, viscous force, and external force, respectively. External force consists of surface tension, gravity, and volume force, where F_{st}, ρg, and F_{vf} represent surface tension, gravity, and volume force, respectively. In addition, in this multi-physics field coupled model, the laminar flow field is coupled to the electrostatic field by applying volume forces to the Navier–Stokes equations, and the electrostatic field can be obtained by calculating the divergence of the Maxwell stress tensor [32]. In the establishment of the two-dimensional EWD simulation model, the Maxwell stress tensor is shown in Equation (10).

$$T = \begin{bmatrix} T_{xx} & T_{xy} \\ T_{yx} & T_{yy} \end{bmatrix} \tag{10}$$

Since the laminar flow change caused by the electrostatic field is reflected by the volume force, the expression of the volume force can be obtained as shown in Equation (11).

$$F = \begin{bmatrix} \frac{\partial(T_{xx})}{\partial x} & \frac{\partial(T_{xy})}{\partial y} \\ \frac{\partial(T_{yx})}{\partial x} & \frac{\partial(T_{yy})}{\partial y} \end{bmatrix} \tag{11}$$

2.2.2. Boundary Conditions

In this multi-physics field coupling model, it is necessary to determine the corresponding boundary conditions to obtain the solution for the governing equations. For the boundary conditions of the electrostatic field, the top of the model is grounded, the bottom outer boundary is connected to the designed pulse, and the remaining boundaries are assigned to zero charge. For laminar boundary conditions, the liquid–solid interface is defined as a wet wall condition, the initial values of pressure and velocity in the laminar field are set to 0, and the wall condition is set to Navier slip. At the same time, the interface of two-phase flow is selected as the initial boundary condition, and both sides of the model (except the pixel wall) are selected as the exit conditions.

2.2.3. Model Simulation

A two-dimensional simulation model was established, based on Table 1, and its model is shown in Figure 3b. The water and oil in the model were set to incompressible flow. Assuming that the temperature of the fluid remains constant, the thermal expansion of the fluid can be ignored and the effect of pressure on dynamic viscosity can be ignored.

Based on the aforementioned model, simulation verification was conducted for the designed pulse. As shown in Figure 3a, this pulse incorporates a direct linear function between the overdriving voltage and the target voltage, which can effectively mitigate abrupt voltage changes and achieve the effect of eliminating glitches. In addition, in order to verify whether the designed pulse in this paper had the effect of eliminating glitches, model simulation verification and comparison were carried out first. The designed linear switching pulse and step switching pulse were introduced into the designed simulation model, as shown in Figure 3c. In the figure, the solid blue line represents the designed pulse, and the dashed red line represents the step switching pulse. Both the designed pulse and step switching pulse have an overdriving voltage of 22 V and a target voltage of 20 V. Additionally, the overdriving time and switching time of the designed pulse are both set to 2 ms, and the overdriving time of the step pulse is set to 4 ms. Within the first 2 ms, the oil shrinks rapidly to one side of the model. Then, with the increase in time, since the voltage of the step switching model remains constant, the oil contraction state remains unchanged, while the driving voltage of the designed linear switching model decreases linearly, resulting in an increase in contact angle so that the oil has a backflow phenomenon, but this backflow phenomenon is not obvious. Figure 3d shows the oil contraction behavior under different pulse modes at 3 ms. The region between the white arrows shows the differences in oil contraction state between the step and linear switching pulses, which arise from their different applied voltage strategies. With the step switching pulse, the voltage remains constant, and the oil contraction state remains essentially unchanged; in contrast, under the linear switching pulse, the voltage decreases linearly over time, causing the slight backflow of the oil. In Figure 3e, the region between the white arrows shows that, at 4 ms and 5 ms, the step switching pulse induces significant backflow due to the voltage abrupt change, which manifests as a sharp glitch on the luminance curve. In contrast, the region between the white arrows in Figure 3f indicates that under the linear switching pulse, the voltage decreases linearly, resulting in a much gentler oil recession process; when the voltage reaches the target value, the oil contraction state stabilizes, thereby effectively suppressing glitches in the luminance curve. Figure 3g shows the dynamic response of the simulation model within the critical time range of 0 ms to 5 ms after the application of the linear pulse voltage. During the overdrive phase (0–2 ms), the relatively high driving voltage causes the oil film to contract rapidly. Subsequently, during the linear switching phase (2–4 ms), as the driving voltage decreases smoothly, the oil–water interface undergoes a smooth and continuous contraction under the combined influence of surface tension and the electric field force. No significant backflow is observed, indicating that the dynamic behavior during this phase is effectively controlled. Finally, during the steady voltage driving phase (4–5 ms), the oil stabilizes near the target position. Therefore, the simulation results show that the designed pulse can effectively eliminate the glitch phenomena in the luminance curve, verifying its effectiveness in the theoretical model. To further confirm the practical performance of this pulse, an experimental platform was built for experimental verification and is detailed in the next section.

Table 1. EWD 2D simulation model parameters.

Parameters	Quantity	Value	Unit
Material	Density of oil	763	kg/m^3
	Density of water	998	kg/m^3
	Dynamic viscosity of oil	2×10^{-3}	Pa·s
	Dynamic viscosity of water	1.01×10^{-3}	Pa·s
	Dielectric constant of oil	4	1
	Dielectric constant of water	80	1
	Dielectric constant of hydrophobic insulating layer	1.934	1
	Dielectric constant of pixel wall	3.28	1
Structure	Width of pixel	100	μm
	Height of pixel wall	3.2	μm
	Width of pixel wall	10	μm
	Thickness of hydrophobic insulating layer	0.4	μm
	Thickness of oil	3.2	μm
Interfacial	Surface tension of oil	0.02	N/m
	Contact angle at top of pixel wall	100	deg
	Contact angle on side of pixel wall	64.8	deg
	Contact angle of hydrophobic insulating layer	35	deg

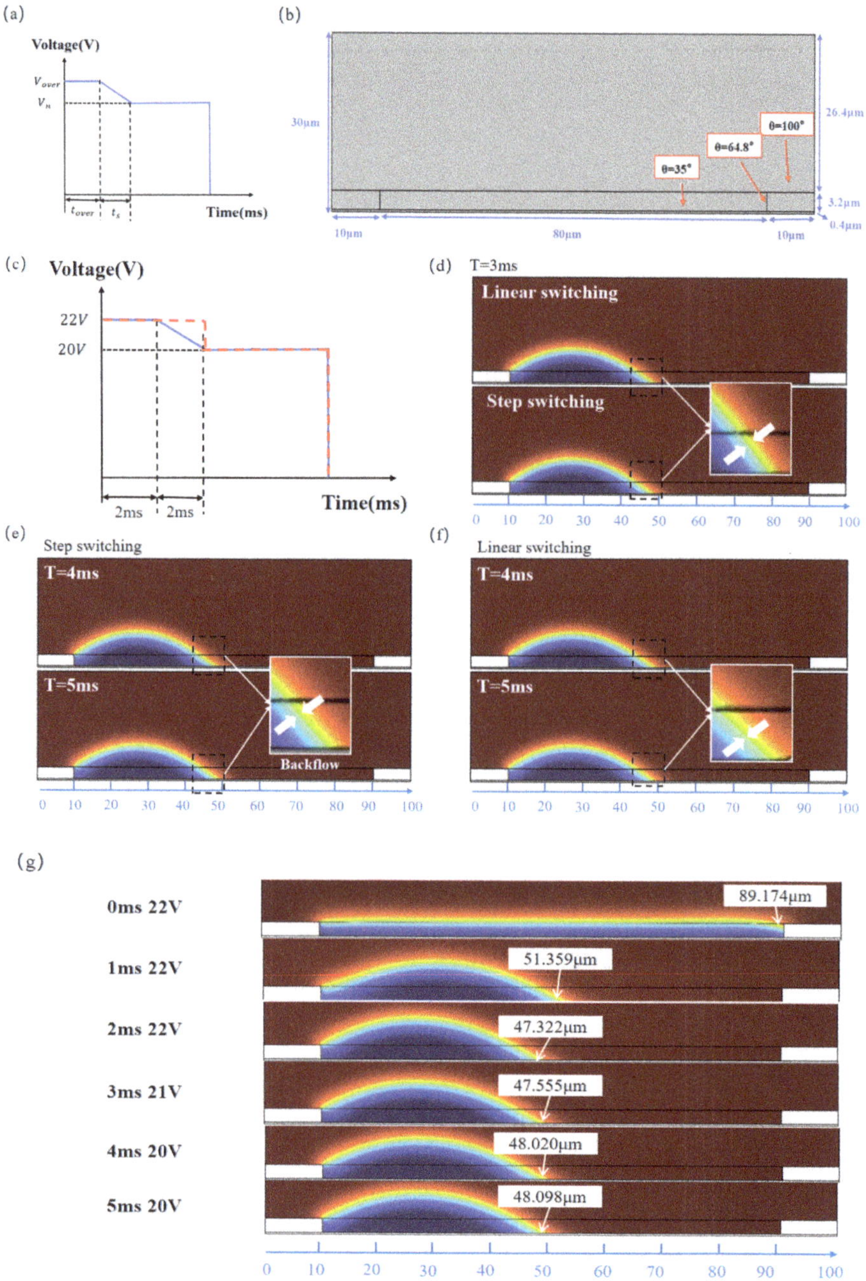

Figure 3. The 2D geometric model and simulation comparison. (**a**) The designed pulse, where t_s represents the switching phase time, V_n represents the stable driving voltage value, and V_{over} and t_{over} represent the overdrive voltage and time, respectively; (**b**) the 2D geometric model; (**c**) a comparison of step and linear switching pulse; (**d**) the oil contraction behavior under different pulses at 3 ms; (**e**) the oil contraction behavior under the step switching pulse at 4 ms and 5 ms; (**f**) the oil contraction behavior under the linear switching pulse at 4 ms and 5 ms; (**g**) simulation of oil contraction under the proposed linear driving pulse from 0 ms to 5 ms.

3. Experimental Results and Discussion

3.1. Experimental Platform

In order to test the validity of the pulse, an experimental platform [39,40] was developed. The platform consisted of an input system and a test system. The input system consisted of a computer, a waveform generator for inputting the pulse, and a signal amplifier for amplifying the pulse signal. The test system consisted of a computer and a colorimeter for testing the response time of EWD.

An EWD was used as the test object in this experiment. The manufacturing process parameters of the EWD are shown in Table 2. In the test process, the pulse was designed and compiled in Matlab (Version R2024A) on the computer, and then the pulse was edited by the Arbexpress waveform editing software (Version 3.4). Then, the edited pulse was imported into the function generator through a universal serial bus (USB) interface, amplified by the signal amplifier, and measured by a colorimeter. Finally, the luminance data was recorded in real time by Admesy software (Version 2.9.3).

Table 2. EWD manufacturing process parameters.

Substrate	Hydrophobic Insulating Layer Thickness	FP Curing	Activation	Pixel Wall Height	Oil Color	Descum	Backflow
0.55 mm UP ITO	400 nm	185 °C 30 min	Power 10 W Time 6 s	3.5 µm	Magenta (R21 5%)	Power 100 W Time 100 s	200 °C/1 h

3.2. Pulse Design

Figure 4 shows the designed driving pulse for mitigating glitch phenomena. The proposed pulse consists of an overdriving phase, a switching phase, and a driving phase. In the overdriving phase, an overdriving voltage is applied to improve the response speed of the oil, and this duration is called the overdriving time. In the switching phase, the linear function is designed to prevent abrupt voltage changes and abrupt luminance changes, so as to optimize the phenomenon of luminance instability. During the driving phase, a target voltage of 20 V was applied to ensure that the EWD was not damaged during the experiment. Charge trapping affects both the response time and luminance of the EWD, and linear switching can slow down capacitive discharge to eliminate glitches. Therefore, it is necessary to experimentally determine the overdriving voltage, overdriving time, and linear switching time to achieve optimal performance.

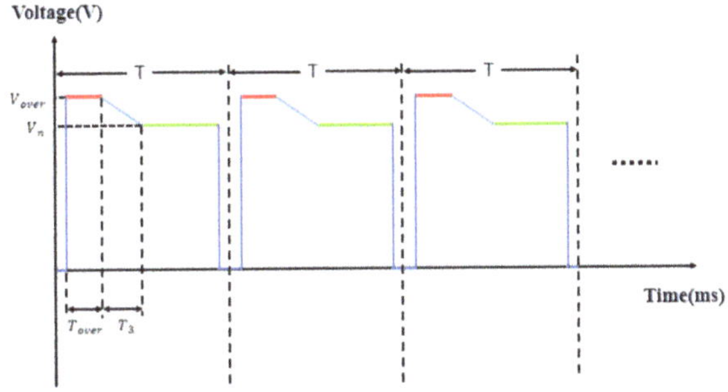

Figure 4. The proposed EWD driving pulse, which consists of an overdriving phase (bold red section), a switching phase (bold blue section), and a driving phase (bold green section), where V_{over} and T_{over} represent the overdriving voltage and the overdriving time, respectively, T_3 represents the switching time, and V_n represents the steady driving voltage.

3.3. *Overdriving Phase Testing*

For the designed pulse, this experiment initially sets the switching phase duration to 2 ms and uses a colorimeter to measure the luminance rise curve of the EWD. This approach was adopted to analyze the effects of overdriving voltage and duration on luminance response characteristics and glitch phenomena. When analyzing the effects of overdriving voltage, overdriving voltages were set to 22 V, 24 V, 26 V, 28 V, and 30 V, with a constant duration of 4 ms, so as to prevent experimental errors caused by structural damage to partial pixels under excessively high voltage [30]. The luminance curves of different overdriving voltages are shown in Figure 5a. It can be seen that when the applied voltage is lower than the threshold voltage of 10 V, the EWD is in the "off state", corresponding to a luminance level of 210. Once the applied voltage exceeds 10 V, the oil begins to retract. The response time is defined as the time interval from the initial moment until the luminance reaches its average steady-state level, where the average steady-state luminance is calculated as the arithmetic mean of the last 30 data points in each voltage column. Meanwhile, the difference between the maximum luminance value when the glitch phenomenon occurs and the minimum luminance value it drops to afterward is defined as the glitch amplitude. When the glitch amplitude is less than 2, it can be regarded as normal luminance curve fluctuation. The experimental results show that as the overdriving voltage increases, the rising speed of luminance first accelerates and then slows down. When the voltage remains below 26 V, raising the voltage accelerates the response; however, once the voltage exceeds 26 V, the rising speed decreases. The reason for this phenomenon is that when the overdriving voltage is higher than 26 V, it drives more ions to migrate to the hydrophobic insulating layer and induce charge trapping. These trapped charges form a reverse electric potential, leading to a decrease in the effective driving voltage. At the same time, when the overdriving voltage is suddenly switched to the target voltage, the charges stored in the capacitor will be rapidly discharged through the circuit, causing the luminance to decrease. This sudden change forces the oil to shrink rapidly, the luminance rises briefly and then decreases as the electric field weakens, forming glitch phenomena. It can be seen from Figure 5b,d that when the overdriving voltage is 22 V the curve is smooth, without obvious fluctuation, and the glitch amplitude is the smallest; when the overdriving voltage is greater than or equal to 24 V, the glitch amplitude of luminance is more obvious. In addition, it can be observed from Figure 5c that as the overdriving voltage increases, the stabilized luminance value decreases. Therefore, increasing the overdriving voltage will lead to an increase in glitch amplitude, an extension of response time, and a decrease in stabilized luminance. When the overdriving voltage is 22 V, the response time is the shortest, 65.55 ms, the luminance is the highest, 930.51, and the glitch amplitude is the lowest, 0.721. Therefore, the overdriving voltage was set to 22 V for a better display effect.

After determining the optimal parameters of the overdriving voltage, the influence of the overdriving time on luminance and the glitch phenomenon is further analyzed by testing the luminance–time curve. The overdriving voltage was set to 22 V, and the overdriving time was set to 2 ms, 4 ms, 6 ms, 8 ms, and 10 ms, respectively. The luminance rise curves corresponding to different overdriving times are shown in Figure 6a. At first, with the increase in overdriving time, the luminance rises faster and faster, and the luminance increases gradually. However, when the overdriving time increases to a certain value and continues to increase, the luminance rise speed decreases and glitches appear, as shown in Figure 6b. The reason for this phenomenon is that the excessive overdriving time (>6 ms) increases the time maintaining the reverse electric field, resulting in more accumulated charges in the hydrophobic insulating layer, reducing the effective driving voltage greatly. Meanwhile, the stabilized luminance decreases and the glitch amplitude increases, as shown in Figure 6d. Meanwhile, as shown in Figure 6c, the glitch amplitude is approximately 0 when the overdriving time is 2 ms or 4 ms, and the luminance value is the highest at 930.51 when the overdriving time is 4 ms. Therefore, the overdriving time was set to 4 ms for a better display effect.

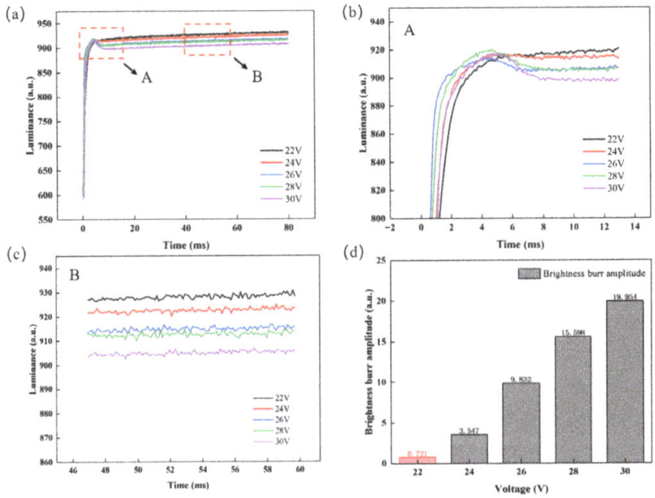

Figure 5. Test data for different overdriving voltages. (**a**) Luminance–time curves at different overdriving voltages, where the response times corresponding to overdriving voltages of 22 V, 24 V, 26 V, 28 V, and 30 V are calculated to be 65.55 ms, 66.93 ms, 68.27 ms, 69.34 ms, and 71.29 ms, respectively; (**b**) enlarged view of area A in Figure 5a; (**c**) enlarged view of area B in Figure 5a; (**d**) comparison of luminance glitch amplitudes under different overdriving voltages.

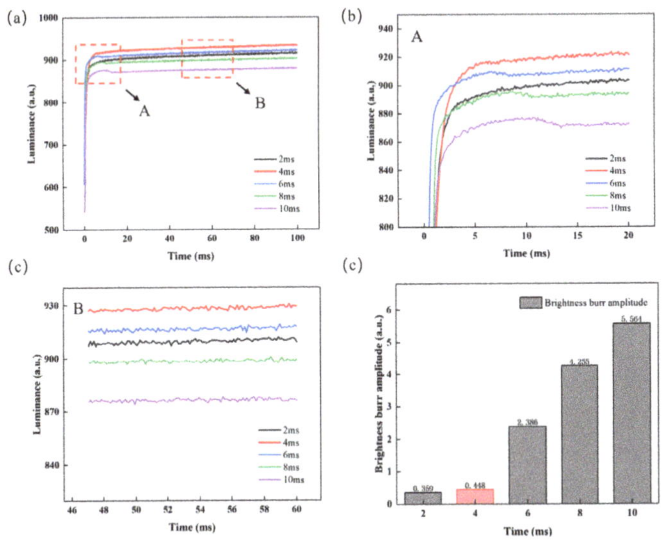

Figure 6. Test data for different overdriving times. (**a**) Luminance–time curves for different overdriving times, where the response times corresponding to overdriving times of 2 ms, 4 ms, 6 ms, 8 ms, and 10 ms are calculated to be 67.5 ms, 65.55 ms, 67 ms, 68.5 ms, and 70.12 ms, respectively; (**b**) enlarged view of area A in Figure 6a; (**c**) enlarged view of area B in Figure 6a; (**d**) luminance glitch amplitude contrast diagram at different overdriving times.

3.4. Test for Switching Phases

In the test experiments for the overdriving phase, it is concluded that when the overdriving voltage is 22 V and the overdriving time is 4 ms, the glitch amplitude value of the luminance curve is already less than 2, which can be regarded as normal fluctuation. To further improve display performance metrics such as response speed and luminance, the timing of the linear switching phase was investigated. This phase uses linear function switching to eliminate the glitch phenomenon. The voltage and time of the overdriving phase were set to 22 V and 4 ms, respectively, and the switching phase time was set to 0.5 ms, 1 ms, 1.5 ms, 2 ms, and 2.5 ms, respectively, for testing. The luminance rise curves under different switching times are shown in Figure 7a. It is observed that the average glitch amplitude in the luminance curve remains below 2, while the response time shows almost no change as the switching time increases. As the switching time increases, the slope of the linear function decreases gradually, leading to a reduction in switching speed. In Figure 7b, when the switching time is less than 2 ms, the luminance slightly decreases as the switching time increases. This is primarily because longer switching times lead to increased charge trapping, accompanied by capacitive discharge effects. However, the linearly decreasing voltage effectively mitigates abrupt voltage changes, thereby suppressing the rapid discharge of the capacitor, resulting in minimal changes in luminance. When the switching time is greater than 2 ms, the response time is extended, and the luminance level is relatively low. Meanwhile, when the switching time is 2 ms, the highest luminance is obtained, which is 930.51. Therefore, the switching time was set to 2 ms.

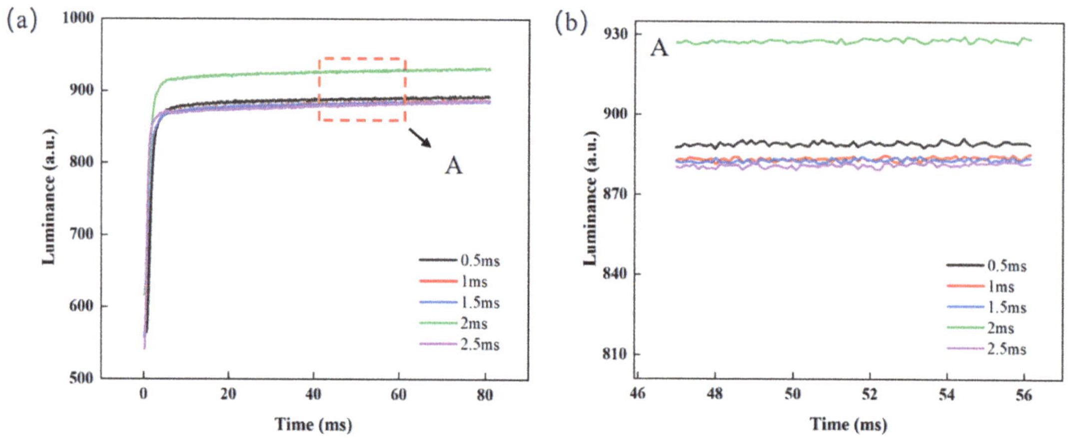

Figure 7. Test data for different switching times. (**a**) Luminance–time curves at different switching times, where the response times corresponding to switching times of 0.5 ms, 1 ms, 1.5 ms, 2 ms, and 2.5 ms are calculated to be 66.2 ms, 65.8 ms, 66 ms, 65.55 ms, and 66.5 ms, respectively; (**b**) enlarged view of area A in Figure 7a.

3.5. Performance Comparison

According to the experimental results, the overdriving voltage, the overdriving time, and the linear switching time of the designed pulse are set to 22 V, 4 ms, and 2 ms, respectively. In order to prove the effect of the designed pulse, it was compared to the PWM pulse [41,42] and the step switching pulse [21]; the PWM pulse and the step switching pulse are shown in Figure 8a. The voltage amplitude of the PWM pulse is 20 V, and the duty ratio is 80%. The overdriving voltage and the overdriving time of the step switching pulse are 22 V and 6 ms, respectively, and the target voltage is 22 V. The luminance rise curves of the proposed pulse, the PWM pulse, and the step switching pulse are shown in Figure 8b. It can be seen that the response time of the proposed pulse is shorter than that of the step switching pulse, the luminance is higher, and the glitch amplitude is effectively reduced. Specifically, as can be seen from Figure 8c, the glitch amplitude of the proposed pulse is less than 2, while the glitch amplitude of the step switching pulse is 2.347. This indicates that the proposed pulse effectively suppresses the luminance glitches caused by the overdriving voltage. Meanwhile, the response time of the proposed pulse is 65.55 ms, the response time of the step switching pulse is 66.77 ms, and the response time of the PWM pulse is 71.29 ms. Therefore, the response time is shortened by 1.82% compared with the step switching pulse and by 8.05% compared with the PWM pulse. The luminance curves of the proposed pulse, the PWM pulse, and the step switching pulse after their input stabilizes are shown in Figure 8d. The steady luminance of the proposed pulse is 930.51, the steady luminance of the step switching pulse is 921.04, and the steady luminance of the PWM pulse is 912.65. Compared with the step switching pulse and the PWM pulse, the achievable luminance is increased by 1.02% and 1.96%, respectively. Figure 8e and Table 3 show the photographs of the instantaneous display states at the selected positions during the glitch occurrence phase (area A) and the stable phase (area B), along with their corresponding aperture ratio values. The PWM pulse and the proposed pulse maintain a constant aperture ratio in both the glitch occurrence phase and the stable phase, while the step switching pulse has a larger aperture ratio in the glitch occurrence phase than in the stable phase. It can be observed that the proposed pulse achieves a uniform and stable transition from the overdriving voltage to the driving voltage, and compared with the two reference driving methods, it realizes higher luminance and superior stability in the stable phase. In EWDs, the step switching of the overdriving voltage causes an abrupt change in the electric field, leading to the rapid discharge of the equivalent capacitor and the generation of a reverse electric field, thereby inducing glitch phenomenon. In this study, by optimizing the overdriving pulse, transient changes in the electric field are mitigated, achieving the goal of eliminating glitches. Meanwhile, compared with the step switching pulse and the PWM pulse, the proposed pulse improves both luminance and response speed.

Table 3. The aperture ratios in different regions corresponding to different pulses.

	PWM Pulse	Step Switching Pulse	Linear Switching Pulse
The aperture ratio of area A (%).	60.42	62.25	61.48
The aperture ratio of area B (%).	60.39	60.07	61.45

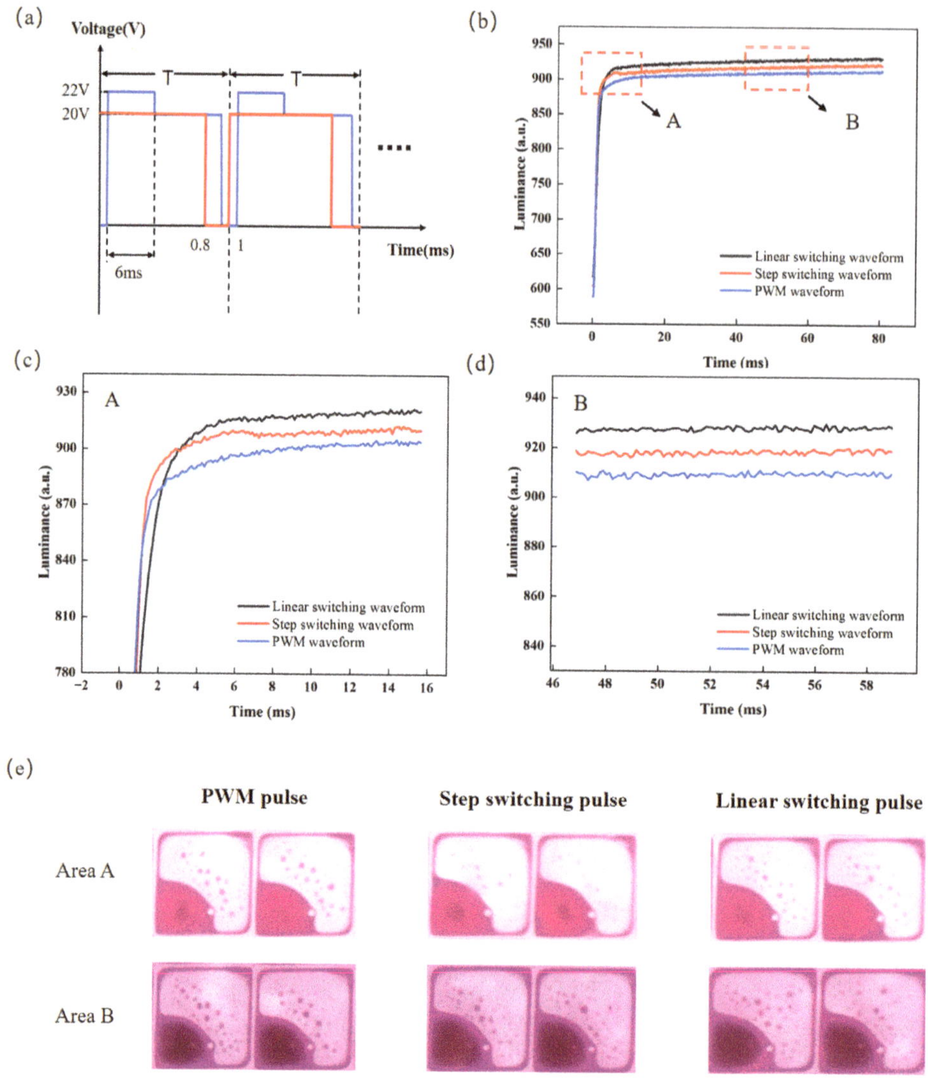

Figure 8. A comparison among the linear switching pulse, step switching pulse, and PWM pulse. (**a**) PWM pulse and step switching pulse. (**b**) Luminance–time curves under the proposed pulse, step switching pulse, and PWM pulse. (**c**) Enlarged view of area A in Figure (**b**). (**d**) Enlarged view of area B in Figure (**b**). (**e**) Photographs of the instantaneous displays of the EWD states corresponding to the two areas in Figure 8b: (top row) the display state during the glitch occurrence phase (area A); (bottom row) the display state during the stable phase (area B). The corresponding driving pulses are as follows: (left) PWM pulse, (middle) step switching pulse, (right) linear switching pulse proposed in this paper.

4. Conclusions

Aiming at the glitch phenomenon in EWDs under overdriving voltages, a multi-stage driving pulse strategy is proposed. By integrating three stages, namely overdriving, switching, and driving, this method effectively ensures the reliability of the device while improving the response performance. Meanwhile, combined with electrowetting theory and equivalent circuit modeling, a quantitative relationship between the driving voltage and the thickness of the insulating layer is established, which provides a basis for the safe selection of the target voltage. Moreover, the results of multi-physics simulation show that changing traditional step switching to linear gradient control can significantly suppress oil backflow and sudden changes in the electric field. This linear transition mechanism effectively delays the capacitive discharge process and reduces the accumulation of charge trapping effects, thereby inhibiting the formation of reverse electric fields and significantly reducing luminance fluctuations. The proposed driving strategy not only realizes glitch-free display but also significantly improves optical response speed and stability, providing a reliable solution for the design of driving systems for high-performance EWDs.

Author Contributions: Conceptualization, Y.Y. and Z.Y.; methodology, Y.Y.; simulation, Y.Y., Q.L. (Qingsong Lu), J.W., and W.X.; experimental platform provision, Y.Y.; Z.Y.; F.C., and L.L.; experiment and data collation, Y.Y., Q.L. (Qingsong Lu), Q.L. (Qifu Liu), and W.X.; writing—original draft preparation, Y.Y.; writing—review and editing, Y.Y. and Z.Y. All authors have read and agreed to the published version of the manuscript.

Funding: This research is funded by the National Natural Science Foundation of China (NO. 62575059), the Innovation Team Project of Regular Universities in Guangdong Province (2024KCXTD056), the Key Laboratory of Regular Universities in Guangdong Province (2023KSYS011), and the Education Planning Leadership Group Teaching and Research Project in Guangdong Province (2022GXJK111).

Institutional Review Board Statement: Not applicable.

Informed Consent Statement: Not applicable.

Data Availability Statement: Data are contained within the article.

Conflicts of Interest: The authors declare no conflicts of interest.

Abbreviation

The following abbreviation is used in this manuscript:

EWD	Electrowetting display

References

1. Sureshkumar, P.; Bhattacharyya, S.S. Display Applications of Electrowetting. *J. Adhes. Sci. Technol.* **2012**, *26*, 1947–1963. [CrossRef]
2. Fan, S.K.; Chiu, C.P.; Huang, P.W. Transmittance tuning by particle chain polarization in electrowetting-driven droplets. *Biomicrofluidics* **2010**, *4*, 043011. [CrossRef] [PubMed]
3. Kim, D.Y.; Steckl, A.J. Electrowetting on Paper for Electronic Paper Display. *Acs Appl. Mater. Interfaces* **2010**, *2*, 3318–3323. [CrossRef] [PubMed]
4. Hayes, R.A.; Feenstra, B.J. Video-speed electronic paper based on electrowetting. *Nature* **2003**, *425*, 383–385. [CrossRef]
5. Chiu, Y.H.; Liang, C.C.; Chen, Y.C.; Lee, W.Y.; Chen, H.Y.; Wu, S.H. Accurate-gray-level and quick-response driving methods for high-performance electrowetting displays. *J. Soc. Inf. Disp.* **2011**, *19*, 741–748. [CrossRef]
6. Tian, L.X.; Bai, P.F. A Combined Pulse Driving Waveform with Rising Gradient for Improving the Aperture Ratio of Electrowetting Displays. *Front. Phys.* **2021**, *9*, 709151. [CrossRef]
7. Luo, Z.J.; Fan, J.J.; Xu, J.Z.; Zhou, G.F.; Liu, S.Y. A novel driving scheme for oil-splitting suppression in Electrowetting display. *Opt. Rev.* **2020**, *27*, 339–345. [CrossRef]
8. Luo, Z.J.; Peng, C.L.; Liu, Y.J.; Liu, B.Q.; Zhou, G.F.; Liu, S.Y.; Chen, N.X. A Low-Cost Drive and Detection Scheme for Electrowetting Display. *Processes* **2023**, *11*, 586. [CrossRef]
9. Zhang, T.Y.; Wang, L.; Liu, L.W.; Li, W.; Wu, S.P.; Guo, J.Y.; Zhou, G.F. Optimizing Bipolar Reset Waveform to Improve Grayscale Stability in Active Matrix Electrowetting Displays. *Micromachines* **2024**, *15*, 1247. [CrossRef]
10. Li, W.; Wang, L.; Zhou, G.F. Driving Waveform Design for Quick Response Electrowetting Displays Based on Asymmetric Pulse Width Modulation. *J. Nanoelectron. Optoelectron.* **2020**, *15*, 1293–1299.

11. Jiang, C.D.; Tang, B.; Xu, B.J.; Groenewold, J.; Zhou, G.F. Oil Conductivity, Electric-Field-Induced Interfacial Charge Effects, and Their Influence on the Electro-Optical Response of Electrowetting Display Devices. *Micromachines* **2020**, *11*, 702. [CrossRef] [PubMed]
12. Liu, L.W.; Wu, Z.Y.; Wang, L.; Zhang, T.Y.; Li, W.; Lai, S.F.; Bai, P.F. Design of an AC Driving Waveform Based on Characteristics of Electrowetting Stability for Electrowetting Displays. *Front. Phys.* **2020**, *8*, 618752. [CrossRef]
13. Li, S.X.; Xu, Y.J.; Zhan, Z.Y.; Du, P.Y.; Liu, L.W.; Li, Z.K.; Wang, H.W.; Bai, P.F. Dynamic Adaptive Display System for Electrowetting Displays Based on Alternating Current and Direct Current. *Micromachines* **2022**, *13*, 1791. [CrossRef]
14. Zhang, T.Y.; Deng, Y. Driving Waveform Design of Electrowetting Displays Based on a Reset Signal for Suppressing Charge Trapping Effect. *Front. Phys.* **2021**, *9*, 672541. [CrossRef]
15. Lai, S.F.; Zhong, Q.H.; Sun, H.L. Driving Waveform Optimization by Simulation and Numerical Analysis for Suppressing Oil-Splitting in Electrowetting Displays. *Front. Phys.* **2021**, *9*, 720515. [CrossRef]
16. Riahi, M.; Brakke, K.A.; Alizadeh, E.; Shahroosvand, H. Fabrication and characterization of an electrowetting display based on the wetting-dewetting in a cubic structure. *Optik* **2016**, *127*, 2703–2707. [CrossRef]
17. Zhao, Q.; Tang, B.A.; Bai, P.F.; Zhou, R.; Li, H.; Li, F.H.; Groenewold, J.; Zhou, G.F. Dynamic simulation of bistable electrofluidic device based on a combined design of electrode and wettability patterning. *J. Soc. Inf. Disp.* **2018**, *26*, 27–35. [CrossRef]
18. Chae, J.B.; Hong, J.; Lee, S.J.; Chung, S.K. Enhancement of response speed of viscous fluids using overdrive voltage. *Sens. Actuators B-Chem.* **2015**, *209*, 56–60. [CrossRef]
19. Hong, S.J.; Hong, J.; Seo, H.W.; Lee, S.J.; Chung, S.K. Fast Electrically Driven Capillary Rise Using Overdrive Voltage. *Langmuir* **2015**, *31*, 13718–13724. [CrossRef]
20. Supekar, O.D.; Zohrabi, M.; Gopinath, J.T.; Bright, V.M. Enhanced Response Time of Electrowetting Lenses with Shaped Input Voltage Functions. *Langmuir* **2017**, *33*, 4863–4869. [CrossRef] [PubMed]
21. Zeng, W.J.; Yi, Z.C.; Zhao, Y.M.; Zeng, W.B.; Ma, S.M.; Zhou, X.C.; Feng, H.Q.; Liu, L.M.; Shui, L.L.; Zhang, C.F.; et al. Design of Driving Waveform Based on Overdriving Voltage for Shortening Response Time in Electrowetting Displays. *Front. Phys.* **2021**, *9*, 642682. [CrossRef]
22. Liu, L.W.; Bai, P.F.; Yi, Z.C.A.; Zhou, G.F. A Separated Reset Waveform Design for Suppressing Oil Backflow in Active Matrix Electrowetting Displays. *Micromachines* **2021**, *12*, 491. [CrossRef] [PubMed]
23. Tian, L.X.; Li, H. Design Method of Equivalent Driving Waveform Based on Electrowetting Response Characteristics. *Front. Phys.* **2021**, *9*, 730078. [CrossRef]
24. Long, Z.X.; Yi, Z.C.; Zhang, H.; Lv, J.P.; Liu, L.M.; Chi, F.; Shui, L.L.; Zhang, C.F. Toward Suppressing Oil Backflow Based on a Combined Driving Waveform for Electrowetting Displays. *Micromachines* **2022**, *13*, 948. [CrossRef]
25. Tian, L.X.; Lai, S.F.; Zhang, T.Y.; Li, W.; Tang, B.A.; Zhou, G.F. A Multi-Electrode Pixel Structure for Quick-Response Electrowetting Displays. *Micromachines* **2022**, *13*, 1103. [CrossRef]
26. Long, Z.X.; Yi, Z.C.; Zhang, H.; Liu, L.M.; Shui, L.L. Toward Suppressing Charge Trapping Based on a Combined Driving Waveform with an AC Reset Signal for Electro-Fluidic Displays. *Membranes* **2022**, *12*, 1072. [CrossRef]
27. Xu, Y.J.; Li, S.X.; Wang, Z.Y.; Zhang, H.; Li, Z.K.; Xiao, B.; Guo, W.; Liu, L.W.; Bai, P.F. Design of Multi-DC Overdriving Waveform of Electrowetting Displays for Gray Scale Consistency. *Micromachines* **2023**, *14*, 684. [CrossRef]
28. Xu, W.Z.; Yang, H.K.; Luo, D.X.; Wang, L.; Li, L.Y.; Li, X.X.; Zhou, G.F.; Yi, Z.C. Sampling function application in the driving waveform of electrowetting displays toward high performance. *Appl. Opt.* **2025**, *64*, 712–720. [CrossRef]
29. Zhao, Q.; Tang, B.; Dong, B.Q.; Li, H.; Zhou, R.; Guo, Y.Y.; Dou, Y.Y.; Deng, Y.; Groenewold, J.; Henzen, A.V.; et al. Electrowetting on dielectric: Experimental and model study of oil conductivity on rupture voltage. *J. Phys. D-Appl. Phys.* **2018**, *51*, 195102. [CrossRef]
30. Verheijen, H.J.J.; Prins, M.W.J. Reversible electrowetting and trapping of charge: Model and experiments. *Langmuir* **1999**, *15*, 6616–6620. [CrossRef]
31. Jones, T.B.; Fowler, J.D.; Chang, Y.S.; Kim, C.J. Frequency-based relationship of electrowetting and dielectrophoretic liquid microactuation. *Langmuir* **2003**, *19*, 7646–7651. [CrossRef]
32. Lai, S.F.; Tian, L.X.; Shen, S.T.; Yuan, D.; Tang, B. An arc multi-electrode pixel structure for improving the response speed of electrowetting displays. *Front. Phys.* **2022**, *10*, 975317. [CrossRef]
33. Hsieh, W.L.; Lin, C.H.; Lo, K.L.; Lee, K.C.; Cheng, W.Y.; Chen, K.C. 3D electrohydrodynamic simulation of electrowetting displays. *J. Micromech. Microeng.* **2014**, *24*, 125024. [CrossRef]
34. Malikov, Z.M.; Madaliev, M.E.; Chernyshev, S.L.; Ionov, A.A. Validation of a two-fluid turbulence model in comsol multiphysics for the problem of flow around aerodynamic profiles. *Sci. Rep.* **2024**, *14*, 2306. [CrossRef] [PubMed]
35. Liu, F.L.; Cheng, Y.K.; Zhang, J.H.; Tang, B.; Zhou, G.F. Research progress of physics of electrowetting display devices. *Acta Phys. Sin.* **2023**, *72*, 208501. [CrossRef]
36. Yurkiv, V.; Yarin, A.L.; Mashayek, F. Modeling of Droplet Impact onto Polarized and Nonpolarized Dielectric Surfaces. *Langmuir* **2018**, *34*, 10169–10180. [CrossRef]

37. Zhu, G.P.; Yao, J.; Zhang, L.; Sun, H.; Li, A.F.; Shams, B. Investigation of the Dynamic Contact Angle Using a Direct Numerical Simulation Method. *Langmuir* **2016**, *32*, 11736–11744. [CrossRef] [PubMed]
38. Zhou, M.; Zhao, Q.; Tang, B.; Groenewold, J.; Hayes, R.A.; Zhou, G.F. Simplified dynamical model for optical response of electrofluidic displays. *Displays* **2017**, *49*, 26–34. [CrossRef]
39. Wang, L.; Zhang, H.; Li, W.; Li, J.X.; Yi, Z.Y.; Wan, Q.M.; Zhang, J.T.; Ma, P.C. Driving Scheme Optimization for Electrowetting Displays Based on Contact Angle Hysteresis to Achieve Precise Gray-Scales. *Front. Phys.* **2021**, *9*, 655547. [CrossRef]
40. Chen, M.Z.; Lin, S.L.; Mei, T.; Xie, Z.Y.; Lin, J.P.; Lin, Z.X.; Guo, T.L.; Tang, B. Research on Hydrodynamic Characteristics of Electronic Paper Pixels Based on Electrowetting. *Micromachines* **2023**, *14*, 1918. [CrossRef]
41. Zhang, X.M.; Bai, P.F.; Hayes, R.A.; Shui, L.L.; Jin, M.L.; Tang, B.; Zhou, G.F. Novel Driving Methods for Manipulating Oil Motion in Electrofluidic Display Pixels. *J. Disp. Technol.* **2016**, *12*, 200–205. [CrossRef]
42. Oh, J.M.; Ko, S.H.; Kang, K.H. Shape oscillation of a drop in ac electrowetting. *Langmuir* **2008**, *24*, 8379–8386. [CrossRef] [PubMed]

micromachines

Article

Development, Design, and Electrical Performance Simulation of Novel Through-Type 3D Semi Spherical Electrode Detector Based on SOI Substrate

Zhiyu Liu [1,2,†], Tao Long [1,2,†], Zheng Li [1,2,*], Xuran Zhu [1,2], Jun Zhao [1,2], Xinqing Li [2,3], Manwen Liu [4] and Meishan Wang [1,2]

1 School of Integrated Circuits, Ludong University, Yantai 264025, China; 17861822572@163.com (Z.L.); longtao@ldu.edu.cn (T.L.); 18663488310@163.com (X.Z.); zhaojun@ldu.edu.cn (J.Z.); mswang1971@163.com (M.W.)
2 Engineering Research Center of Photodetector Special Chip in Universities of Shandong, Ludong University, Yantai 264025, China; xqli1996zz@163.com
3 School of Materials Science and Engineering, Xiangtan University, Xiangtan 411105, China
4 Institute of Microelectronics, Chinese Academy of Sciences, Beijing 100029, China; liumanwen@ime.ac.cn
* Correspondence: 3636@ldu.edu.cn
† These authors contributed equally to the work.

Abstract: This article proposes a novel three-dimensional trench electrode detector, named the through-type three-dimensional quasi-hemispherical electrode detector. The detector adopts a trench structure to package each independent unit and achieves complete penetration of trench electrodes with the help of an SOI substrate. The horizontal distances from the center anode of the detector to the trench cathode and the detector thickness are equal. It has a near-spherical structure and exhibits spherical-like electrical performance. In this study, we modeled the device physics of the new structure and conducted a systematic three-dimensional simulation of its electrical characteristics, including the electric field, electric potential, electron concentration distribution of the detector, the inducted current caused by incident ions, and the crosstalk between detector units. Computational and technology computer-aided design (TCAD) simulation results show that the detector has an ultra-small capacitance (2.7 fF), low depletion voltage (1.4 V), and uniform electric field distribution. The trench electrodes electrically isolate the pixel units from each other so that the coherence effect between the units is small and can be applied in high-resolution X-ray photon counting detectors to enhance the contrast-to-noise ratio of low-dose imaging and the detection rate of tiny structures, among other things.

Keywords: silicon detector; trench electrode; electrical characteristics; full depletion voltage; SOI

1. Introduction

Silicon detectors are important detectors widely used in fields such as nuclear physics [1], high-energy physics [2], and medical imaging [3]. In the field of nuclear physics, silicon detectors are widely used in particle detection and physical experiments, such as particle identification [4], particle trajectory reconstruction [5], metal element analysis [6], etc. In the field of high-energy physics, silicon detectors are widely used in large-scale collision experiments [7,8] such as CMS [9] and ATLAS [10] for the study of particles such as the Higgs boson [11]. In the medical field, silicon detectors are also widely used in medical imaging technologies such as positron emission tomography (PET) for tumor diagnosis, cardiac lesion detection, etc. [12,13]. Traditional silicon detectors are fabricated using planar technology (two-dimensional detectors), with electrodes being made on the surface of the wafer, resulting in longitudinal depletion. Compared to two-dimensional silicon detectors, the electrodes of three-dimensional detectors are directly fabricated in silicon materials, and the depletion thickness of the detector no longer depends on the wafer thickness. The depletion voltage can be controlled by designing the electrode spacing. Compared to two-dimensional detectors, three-dimensional detectors have superior radiation resistance, more optimized edge design, and more optimized charge collection efficiency. At present, the particle density of the High-Luminosity Large Hadron Collider (HL-LHC) has significantly increased, which has increased the spatial resolution requirements of the detector and promoted the miniaturization of pixel size [14]. At the same time, reducing the electrode spacing becomes key, which can shorten the charge drift path, alleviate charge trapping caused by radiation-induced defects, and enhance radiation hardness. The thickness of detectors tends to be thinner, which can reduce capacitance and adapt to micro electrodes, laying the foundation for performance optimization.

In 1997, Parker et al. from the University of Hawaii proposed the first generation of three-dimensional column electrode silicon detectors [15]. Compared to 2D detectors, the electrode spacing is not limited by the thickness of the chip, resulting in lower full depletion voltage, a shorter charge collection time, better radiation tolerance, and controllable electrode spacing, which improves mechanical stability and radiation resistance [16,17]. However, the detector suffers from a saddle-shaped potential ("dead zone") in the center region of the electrodes and an uneven distribution of the electric field, which is concentrated in the vicinity of the collector electrodes [18]. In 2006, an important milestone was achieved with the invention of "active-edge" sensor technology by Kenney et al. [19]. Their pioneering work solved the key problem of dead zones around the physical boundaries of silicon sensors. Kenney's technology utilizes deep reactive ion etching (DRIE) to create smooth, vertically defined edges. By doping these etched-out trenches and passivating them with a thermal oxide layer, they successfully transformed the mechanical edges into functional sensitive electrodes. This allows the sensitive volume to extend to within a few micrometers of the physical boundary, a significant improvement that is critical for applications where sensors need to be spliced with minimal dead space. However, despite the excellent resolution of the sensor perimeter by the active edge technique, the internal electrode structure and the associated bulk electric field distribution are still based on planar geometries or columnar electrodes in some 3D designs and thus, it is still structurally a planar 2D detector or 3D column electrode detector. These internal structures may still lead to inhomogeneous electric fields, low-field-strength regions, and limitations in further reducing the depletion voltage and readout capacitance [19]. In 2009, Brookhaven National Laboratory (BNL) proposed an innovative three-dimensional trench electrode silicon detector structure [18] which uses a vertical trench structure and a composition of central columnar electrodes. By optimizing the charge collection path, the signal charge collection rate and charge gain performance were significantly improved. Theoretical analysis and experimental results show that the three-dimensional groove electrode structure can achieve uniform electric field distribution inside the detector and effectively reduce the fully depleted operating voltage [20]. However, in order to prevent detachment of the detector cell, some unetched areas will be retained at the bottom, which leads to less utilization of the space inside the detector. In the following years, a breakthrough in the field of edge processing technology for silicon detectors was achieved by the VTT Technical Research Center of Finland. They developed an edge activation process based on ion implantation (instead of the conventional polysilicon filling technique) to prepare edge-free detectors with dead layer thickness below 1 μm on 6-inch wafers [21,22]. The technique achieves edge electrical activation by sidewall ion implantation, which avoids complex steps, such as polysilicon growth and planarization, and significantly improves fabrication efficiency and yield [21,23].

In the recent research on 3D trench detectors, most research institutes have made a lot of effort in the related process and structural design. For example, Forcolin et al. prepared a TIMEPIX-compatible 3D trench electrode pixel sensor for the first time in the framework of the INFN TIMESPOT project using a single-sided process at FBK, and they improved the problem of the uneven electric field of traditional 3D detectors by optimizing the geometry of the trench, which showed that its leakage current was as low as 10 pA/pixel and its capacitance was 70–75 fF/pixel. The optimization of trench geometry improves the problem of uneven electric fields in traditional 3D detectors, and the test shows that the leakage current is as low as 10 pA/pixel, the capacitance of the counter electrode is 70–75 fF/pixel, and the matching test beam experimentally achieves a timing resolution of less than 30 ps. At the same time, it points out that the 13% dead volume of the trench structure and the complexity of the process are the challenges of mass production [24]. In order to deal with the fabrication yields caused by the defects of long trench lithography, Ye et al. proposed an improved 3D trench design by introducing a 10 μm gap in the p^+ ohmic trench and arranging it in a staggered manner with the n^+ readout trench. TCAD simulations and Monte Carlo simulations show that this design has a minimal effect on the uniformity of the electric field and the weighting field, even when the trench is irradiated at 2×10^{16} neq/cm^2, which is the same as that in the p^+ ohmic trench [25]. Liu et al. focused on the irradiation resistance of 3D trench detectors on a N substrate. TCAD simulations show that when the electrode spacing is controlled at 5–20 μm, the full depletion voltage of the detector remains below 500 V and the transient current is lower than 500 V even under extreme irradiation. The TCAD simulation shows that when the electrode spacing is controlled at 5–20 μm, even under extreme irradiation injection, the full depletion voltage of the detector is still lower than 500 V and the peak response time of transient current is <100 ps, which confirms that the small electrode spacing is key to improve the irradiation resistance and provides a basis for the design of applications in a high-radiation environment [26].

In another paper published by our team [27], a new type of detector was designed using a double-sided etching and filling process, which effectively prevents the detector unit from falling off, thus realizing a feed-through electrode design and expanding the internal effective sensitivity area, but the process steps in this way are relatively large. This paper combines this design with the SOI process, also realizes the design of penetrating electrodes, and the whole process uses a single-sided process, which significantly simplifies the process steps; however, the price of SOI wafers is relatively high.

2. Three-Dimensional Semi Spherical Electrode Detector Based on SOI Substrate

Building on the inspiration from the aforementioned detectors, we propose a 3D hemispherical electrode detector based on an SOI substrate, which further innovates upon the 3D electrode structure, representing a true 3D silicon detector. This design, combined with an SOI (silicon-on-insulator) substrate, achieves a quasi-hemispherical electrode design with five sides of cathodes (four sides of doped trench electrodes and one bottom side with an ion implant electrode) surrounding a point-like collection anode for each pixel cell (as shown in Figure 1). It is entirely different from the detector structure presented in [19], where trench walls are present only in the peripheral regions of the detector chip to form an active edge. Our 3D hemispherical electrode detector array is a full 3D detector, whereas [19] uses an active slot wall to surround the entire 2D strip detector. Our new detector offers a more symmetrical electric field distribution than that of the 3D cylindrical trench electrode [20] and 3D column electrode detectors [15]. The small pixel design, where trench electrodes surround a central point-like anode, enables the new detector to have a smaller depletion voltage, smaller readout capacitance, and anti-crosstalk characteristics. The detector unit's cell structure comprises a central point-contact collector electrode and a hemispherical electrode, wherein the hemispherical electrode consists of a peripheral groove electrode (as shown in Figure 1) and an ion-implanted back electrode. The detector achieves 2D position sensitivity through the layout of the array of this unit cell (3D pixel detector). Since cell capacitance is positively correlated with the area and depth of the collection electrode, the point-like collection electrode in our detector can significantly reduce the capacitance of each individual pixel (capacitance modeling, calculations, and simulation validation results will be provided later). Furthermore, such an enclosed design isolates each pixel from others, thus effectively eliminating charge sharing between detector cells. Therefore, this semi-spherical electrode detector has the advantages of low depletion voltage, uniform electric field distribution, small capacitance, and no charge-sharing effect between the detector pixel units, making it suitable for application in high-energy physics (radiation hard) and X-ray detection, such as particle colliders, security check, nuclear safety, and other scenarios. It may meet the requirements for future upgrades and replacements in ATLAS and CMS in the High-Luminosity (HL) Large Hadron Collider.

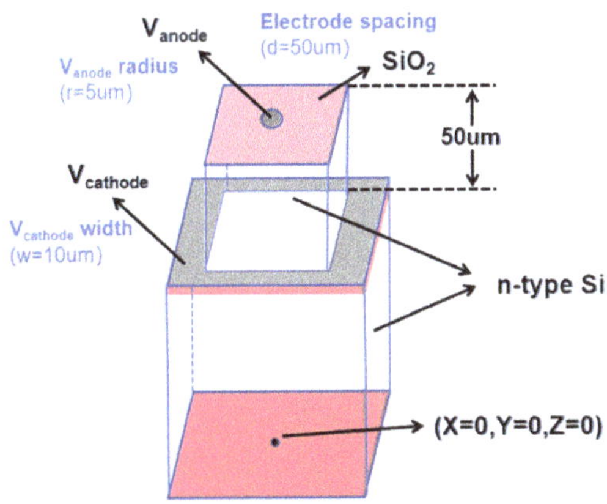

Figure 1. Conceptual diagram of detector.

The detector unit has a rectangular structure, and its silicon substrate is lightly doped n-type with a doping concentration of $1 \times 10^{12}/\text{cm}^3$. Figure 2 shows the schematic diagram of the detector structure, and Figure 2a shows that the pixel units are separated by groove electrodes with a groove width of w. The dot-shaped central anode is arranged at the top of the detector with a radius of r. The vertical distance from the center area of the detector to the groove wall is R. Figure 2b shows that the trench is etched to the P-type heavily doped layer of the SOI substrate, with a trench depth of R. The trench cathode forms the P-type heavily doped layer through an ion implantation process. The cathode of the detector is composed of a trench wall and a P-type heavily doped layer on the substrate, resulting in a distance of R from the center anode of the detector to the trench cathode and the bottom cathode, forming a detector with a structural radius of R. Both the anode and cathode have a thickness of 1 μm and are covered with 1 μm aluminum electrodes on top.

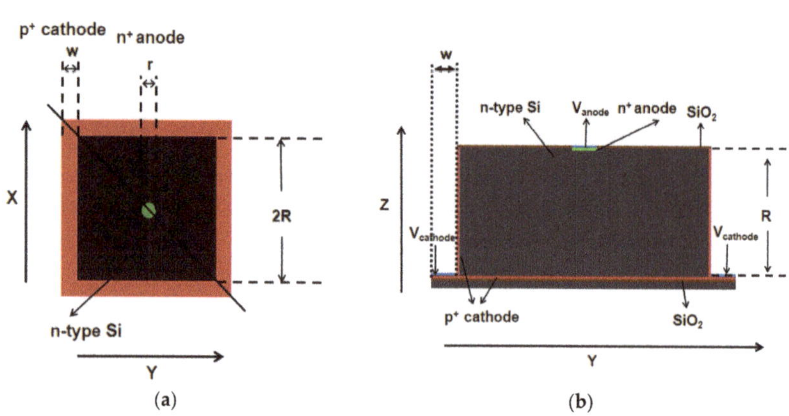

Figure 2. (**a**) Top view; (**b**) sectional view.

Through theoretical calculations, we derive and compute the full depletion voltage and capacitance values of the novel detector. The reliability will be subsequently verified by simulation experiments.

The depletion voltage formula for a hemispherical electrode detector is [28]

$$V_{fd} = \frac{eN_{eff}d^2}{6\varepsilon_r\varepsilon_0} + V_{bi} \quad (1)$$

where q is the charge of each electron, $e = 1.6 \times 10^{-19}$ C; ε_r is the relative dielectric constant of silicon, $\varepsilon_r = 11.9$; ε_0 is the vacuum dielectric constant, $\varepsilon_0 = 8.854 \times 10^{-12}$ F/m; d is the electrode spacing, $d = 50$ μm; N_{eff} is the effective doping concentration of N-type light doping on the silicon substrate, $N_{eff} = 1 \times 10^{12}/\text{cm}^2$; and V_{bi} is the built-in electric potential of the Si unit, $V_{bi} = 0.75$ V. The depletion voltage of the detector can be obtained by using Equation (1), which is 1.4 V.

For hemispherical structures, according to Gauss's theorem:

$$E(r)A(r) = \frac{Q}{\varepsilon_r\varepsilon_0} \quad (2)$$

where $E(r)$ is the electric field at radius r and the surface area of the hemisphere at radius r is

$$A = 2\pi r^2 \quad (3)$$

Therefore, we have

$$E(r) = \frac{Q}{2\pi\varepsilon_r\varepsilon_0 r^2} \quad (4)$$

Therefore, the electric potential difference from the cathode to the anode can be calculated as

$$\Delta U = \int_{r_0}^{R} \frac{Q}{2\pi\varepsilon_r\varepsilon_0}\frac{dR}{r^2} = \frac{Q}{2\pi\varepsilon_r\varepsilon_0}\left(\frac{1}{r_0} - \frac{1}{R}\right) \quad (5)$$

The capacitance of the hemispherical detector is

$$C = \frac{Q}{\Delta U} = \frac{2\pi\varepsilon_r\varepsilon_0 r_0\, R}{R - r_0} \quad (6)$$

where R is the electrode spacing, $R = d = 50$ μm, and r_0 is the center anode radius, $r_0 = 5$ μm. By substituting the size of the detector into Equation (6), the capacitance of the detector can be approximately solved as 3.7 fF.

3. Electrical Characteristics

3.1. Analysis of Electric Field Characteristics

In the field of silicon detector research, the electric field distribution characteristics are one of the key factors in evaluating its performance. By analyzing the two-dimensional cut images obtained under different bias voltage conditions (as shown in Figure 3), we systematically observed the dynamic behavior of the electric field distribution inside the detector with voltage changes. Specifically, the low-electric-field region inside the detector decreases with the gradual increase in the applied bias voltage, and when the bias voltage reaches a certain threshold, i.e., when the detector is completely depleted, the internal electric field covers the whole detector and the zero-electric-field region disappears. When the voltage is −1.4 V or so, the detector is completely depleted, and at this time, there is almost no low-electric field or zero-electric-field region inside the detector. This also means that the carriers can always maintain a directional motion under the action of the electric field force, which enhances the collection efficiency of the carriers and thus improves the response speed and sensitivity of the detector.

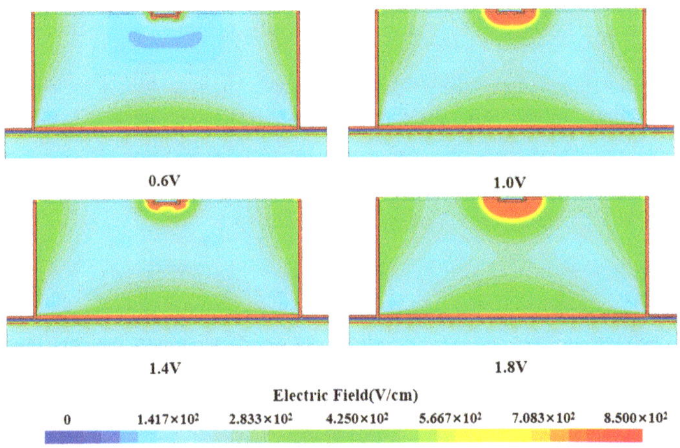

Figure 3. Two-dimensional electric field distribution at X = 0 for three-dimensional quasi-hemispherical electrode detectors at different voltages.

Figure 4 shows a cross-section view of the electric field distribution in the corner region of the detector biased at a voltage of 1.4 V. It can be seen from the figure that the corner region is in a depleted state with a non-zero electric field. Although its electric field strength (at about 100 V/cm) is lower compared to that of the surrounding region, it still has the ability to make carriers to drift for charge collection. In applications where fast charge collection is needed, one can improve the low-electric-field situation by using a hexagonal cell shape for a detector array or a circular one for a single detector, as shown in Figure 5.

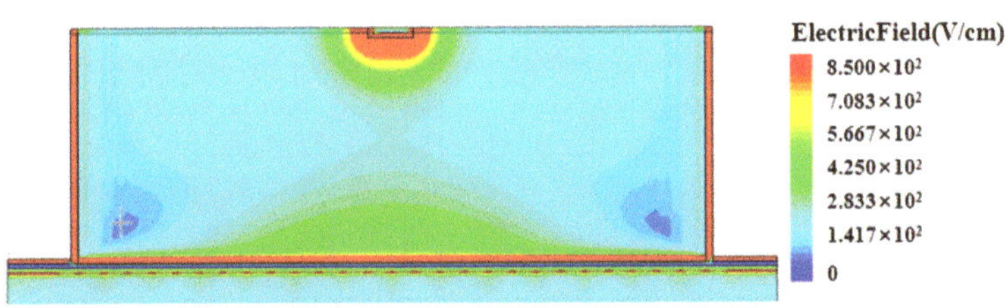

Figure 4. Electric field cross-section at the corner of the detector under 1.4 V voltage (the cut surface is at the diagonal in Figure 2a).

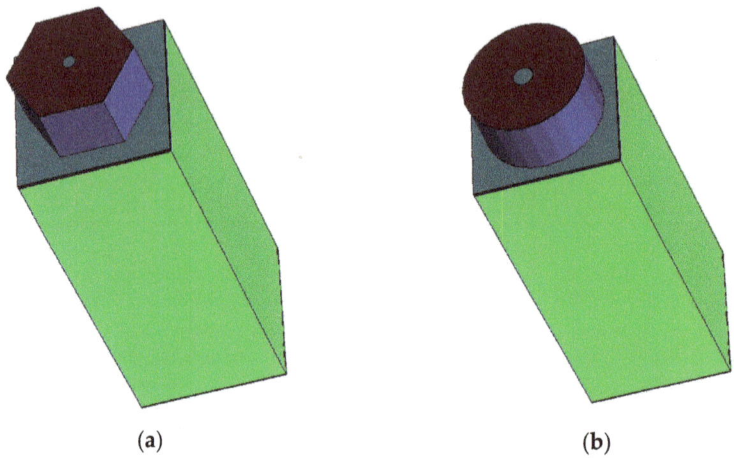

Figure 5. Other unit structures of the detector. (**a**) Hexagonal prism structure; (**b**) cylindrical structure.

3.2. Analysis of Electron Concentration Inside Detector

As shown in Figure 6, this study conducted a systematic simulation analysis of the electric potential distribution characteristics of a three-dimensional quasi-hemispherical detector.

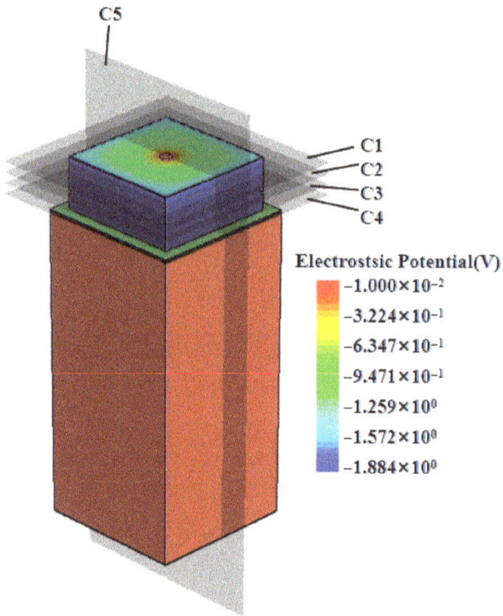

Figure 6. Detector unit potential diagram.

Figure 7 presents the transverse cross-section of the detector's electric potential distribution, revealing a significant ring-shaped characteristic in its internal electric potential (the simulated height is 340 um, the detector substrate height is 290 um, and the detector unit height is 50 um). The potential gradient is smooth and relatively uniform, and the potential shows a concentric-circle symmetric distribution, which means that the detector's electric field in the symmetric direction as well as the motion of the carriers are consistent, improving the overall stability and reliability of the detector and also proving that the collection of incident particles by the detector is not affected by the angle.

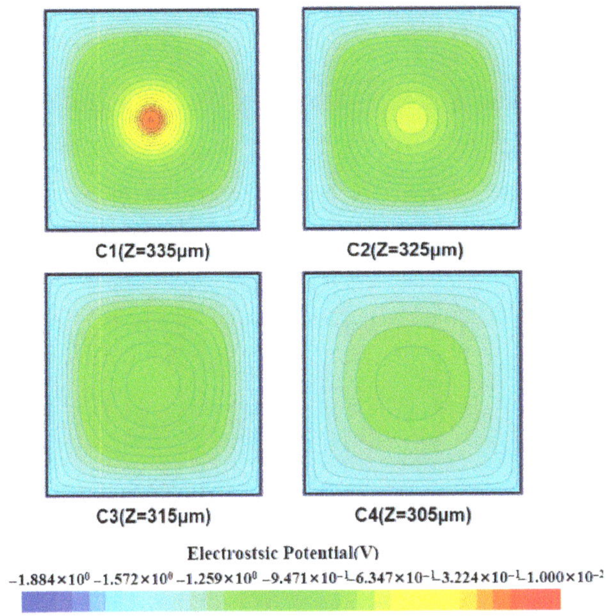

Figure 7. Two-dimensional potential distribution map at different heights of the detector.

3.3. Analysis of Electron Concentration Characteristics

Figure 8 illustrates the distribution of electron concentration inside the detector at different bias voltages. It can be observed from the figure that the electron concentration near the cathode region is relatively low, while the electron concentration near the center collector region is significantly high. As the bias voltage increases, the electron concentration inside the detector decreases gradually. When the bias voltage reaches 1.4 V, the electron concentration inside the detector is already one to two orders of magnitude below its intrinsic doping concentration. And as the bias voltage continues to increase, the electron concentration inside the detector basically no longer changes, a phenomenon that indicates that the detector has reached a state of complete depletion.

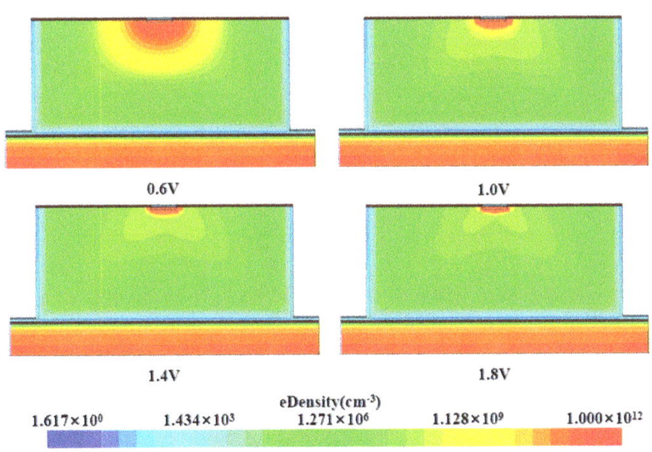

Figure 8. Two-dimensional electron concentration distribution at X = 0 for 3D quasi-hemispherical detectors at different voltages.

As illustrated in Figure 9, the one-dimensional electron density distribution curve is presented as a function of the Z-axis direction. The results demonstrate that, with an increase in the applied voltage, there is a consistent decrease in the electron density within the detector, until it reaches a specific threshold voltage. At this threshold voltage, the electron density within the detector approaches a state of complete exhaustion. When the applied voltage is equal to −1.4 V, the electron concentration in most of the effective sensitive region inside the detector is one to two orders of magnitude lower than the substrate concentration. As the voltage is increased further, the electronic concentration within the effective detection area remains constant, indicating that the detector has reached a state of complete depletion. Consequently, the depletion voltage of the detector is determined to be 1.4 V, which is consistent with the previous theoretical calculation results.

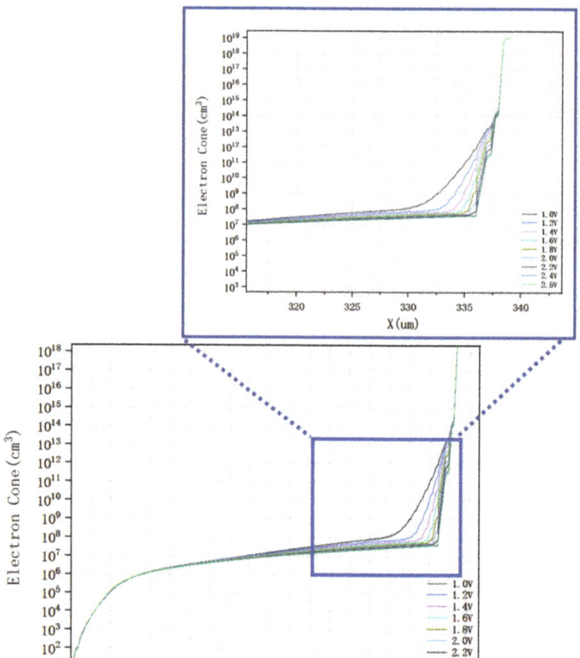

Figure 9. Electron concentration distribution at different voltages.

3.4. Analysis of Transient Induced Current in Detectors and Estimation of Charge Collection Efficiency After Irradiation

As shown in Figure 10, we simulate the capacitance curves of the detector under different irradiation injections by varying the substrate doping concentration of the detector.

When the detector is irradiated by neutrons and charged particles, the effective doping concentration N_{eff} will increase linearly with the 1 MeV neutron-equivalent flux φ_{neq} at high fluxes as (N-type silicon)

$$N_{eff} \cong N_{d0}e^{-\gamma\varphi_{eq}} - \beta\varphi_{eq} \tag{7}$$

where N_{d0} is the initial doping concentration, γ is the donor impurity removal rate, β is the deep acceptor introduction rate, and φ_{eq} is the 1 MeV neutron equivalent flux. Since the γ of proton radiation is about 1×10^{-13} cm^2 and β here is 0.01 /cm [29], in the case of $\varphi_{eq} > 1 \times 10^{14}$ neq/cm^2,

$$N_{eff} \cong -\beta\varphi_{eq} \tag{8}$$

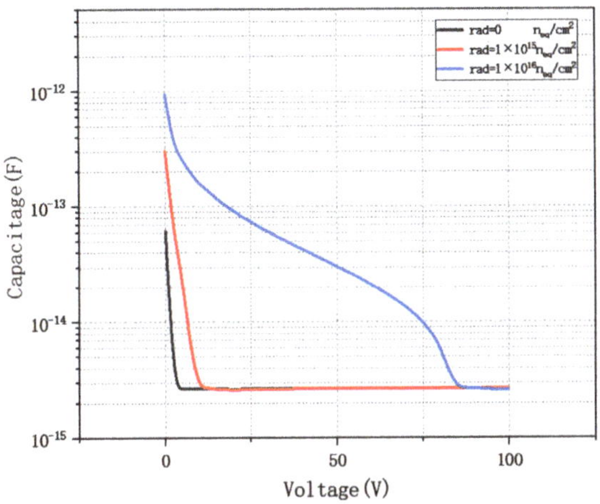

Figure 10. Capacitance curves of the detector at different irradiation injections.

It can be clearly seen from the figure that when the detector reaches the fully depleted state, its capacitance value tends to stabilize and no longer varies with the increase in bias voltage. The capacitance characteristic of the detector is one of the key factors affecting its noise performance. Therefore, the smaller the capacitance, the less the noise. From the simulation results, it can be seen that the depletion voltage of the detector increases as the radiant flux increases. This is because the increase in radiative flux strongly alters the effective doping of the detector substrate [30]. Since the geometrical capacitance only depends on the physical structure of the detector, the saturation capacitance value usually does not change even if the depletion voltage of the detector changes when it is affected by external factors such as radiation. The simulation results further show that the depletion voltage of the detector remains around 88 V when the irradiation flux reaches 1×10^{16} cm^{-2}, which fully verifies the feasibility of the detector to work under a strong irradiation environment. Figure 11 shows the breakdown characteristics of the detector. The breakdown voltage of the detector is 113 V.

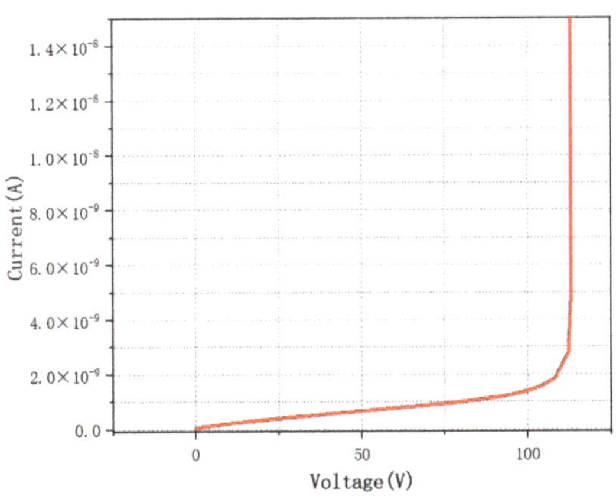

Figure 11. Detector breakdown characteristic diagram.

3.5. Analysis of Transient Induced Current in Detectors

As shown in Figure 12, we simulated the induced current profile of the detector under different irradiation injections by varying the substrate doping concentration of the detector. It can be clearly seen that the induced current of the detector peaks at around 1.6 ns under unirradiated conditions. It is also noted that the peak value of the induced current tends to decrease with increasing irradiation fluence. This phenomenon can be attributed to the damaging effect of irradiation on the detector material. As a result of irradiation, more defects are created inside the detector, which become trapping centers for charge carriers, thus reducing the charge collection efficiency. Below, we can simply estimate the collection time of incident particles for the detector at an operating voltage of 10 V based on the equation.

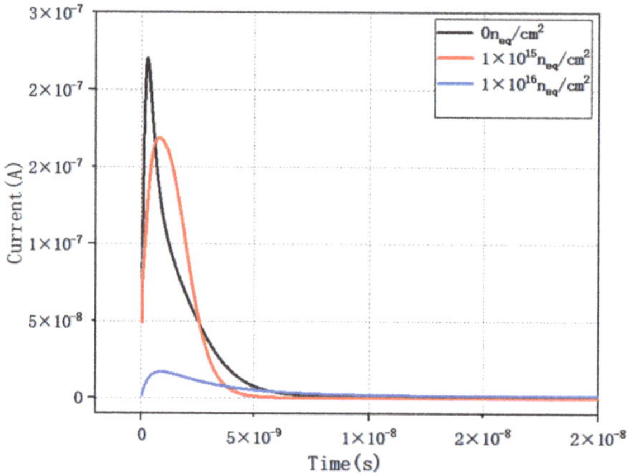

Figure 12. Instantaneous response current under different irradiation conditions.

The average internal electric field of the detector is

$$E = \frac{V}{d} \tag{9}$$

Therefore, the average drift response time of carriers in the detector is

$$t = \frac{d}{\mu E} \tag{10}$$

In TCAD space, the following models were activated in the simulated incident particle collection simulation experiment: DopingDependence, eHighFieldSaturation, hHighField-Saturation, Enormal, and CarrierCarrierScattering. The ion incident point was set to (0, 40, 339). The heavy-ion injection pulse duration was 1×10^{-12} s, with an implantation depth of 40 μm and a linear energy transfer (LET) of 1.28×10^{-5} pC/μm^2. The operating voltage was set to -10 V, and the oxide charge density was converted to 4×10^{11} cm^{-2}.

μ is the electron mobility, $\mu = 1450$ cm^2/(V·S). By substituting the detector size into Eq(10), the induction time is approximately 1.6 ns, which is roughly equivalent to the simulation results in Figure 12.

Figure 13 shows the instantaneous induced current curve of a 3 × 3 detector array. We simulated MIP (minimum ionizing particle) incidence on the lower right unit of the detector (X_3, Y_1) and obtained the induced current. Through observation, we found that no induced current occurred in adjacent units, only in the middle unit with MIP incidents, indicating that there is little crosstalk between detector units.

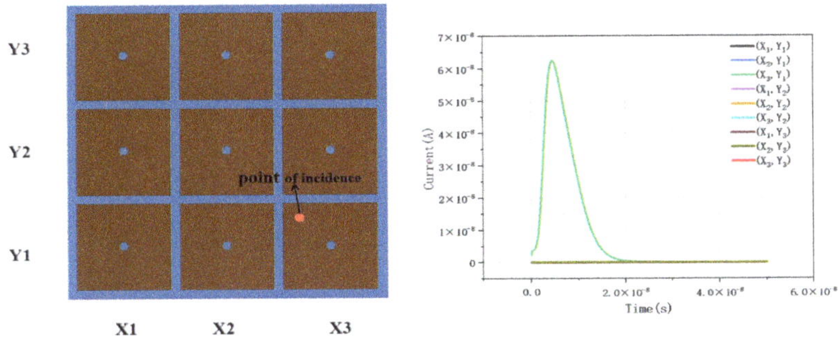

(**a**) (**b**)

Figure 13. Instantaneous response current of incident particles in a 3 × 3 array (**a**) at the incident position and (**b**) response curve.

4. Conclusions

In this study, we present for the first time the structure of a novel detector, named the novel through-type 3D semi-spherical electrode detector based on an SOI substrate. This detector has the advantages of low depletion voltage, ultra-small capacitance, and symmetric distribution of a uniform electric field. In addition, its hemispherical structure ensures that the charge collection is almost independent of the angle θ, thus effectively capturing X-ray signals. It is an ideal X-ray detection tool, capable of effectively detecting hard X-rays and gamma rays, and it thus meets the requirements of high sensitivity and fast response in scenarios such as security check and nuclear safety. It also meets the requirements for future upgrades and replacements in the High-Luminosity (HL) Large Hadron Collider at ATLAS and CMS.

We simulated both individual cells and arrays of detector cells using Senturous TCAD 2019. We obtained the electrical characteristics of the detector such as potential, electric field, electron concentration distribution, capacitance, and incident particle induced current. The following conclusions were drawn:

The new detector utilizes a silicon-on-insulator (SOI) substrate and introduces a feedthrough electrode structure. This design significantly increases the effective sensitivity region of the detector. The full depletion voltage of the detector is measured to be 1.4 V, its geometrical capacitance is about 2.7 fF under non-irradiated conditions, and the response time of the detector is 1.6 ns at a bias voltage of 10 V with an electrode spacing of 50 µm. Simulation analysis shows that the detector array cells are effectively electrically isolated from each other, and the crosstalk can be neglected to ensure the high-energy-resolution performance of the device.

Author Contributions: Conceptualization, Z.L. (Zhiyu Liu) and Z.L. (Zheng Li); methodology, Z.L. (Zheng Li); software, Z.L. (Zhiyu Liu), T.L., and X.Z.; validation, Z.L. (Zhiyu Liu), T.L., J.Z., X.Z., and Z.L. (Zheng Li); formal analysis, Z.L. (Zhiyu Liu); investigation, Z.L. (Zhiyu Liu), T.L., X.L., M.L., and M.W.; resources, Z.L. (Zheng Li); data curation, Z.L. (Zhiyu Liu), T.L., and J.Z.; writing—original draft preparation, Z.L. (Zhiyu Liu); writing—review and editing, Z.L. (Zhiyu Liu) and X.Z.; visualization, Z.L. (Zhiyu Liu); funding acquisition, Z.L. (Zheng Li); supervision, Z.L. (Zheng Li); project administration, Z.L. (Zheng Li). All authors have read and agreed to the published version of the manuscript.

Funding: This research was funded by the National Key R&D Program of China (2023YFF0719600) and the Key Project of National Natural Science Foundation of China (11835008).

Data Availability Statement: The original contributions presented in the study are included in the article, further inquiries can be directed to the corresponding author.

Conflicts of Interest: The authors declare no conflicts of interest.

References

1. He, R.; Niu, X.-Y.; Wang, Y.; Liang, H.-W.; Liu, H.-B.; Tian, Y.; Zhang, H.-L.; Zou, C.-J.; Liu, Z.-Y.; Zhang, Y.-L.; et al. Advances in nuclear detection and readout techniques. *Nucl. Sci. Tech.* **2023**, *34*, 205. [CrossRef]
2. Moser, H.-G. Silicon detector systems in high energy physics. *Prog. Part. Nucl. Phys.* **2009**, *63*, 186–237. [CrossRef]
3. Jiang, W.; Chalich, Y.; Deen, M.J. Sensors for positron emission tomography applications. *Sensors* **2019**, *19*, 5019. [CrossRef]
4. Badalà, A.; La Cognata, M.; Nania, R.; Osipenko, M.; Piantelli, S.; Turrisi, R.; Barion, L.; Capra, S.; Carbone, D.; Carnesecchi, F.; et al. Trends in particle and nuclei identification techniques in nuclear physics experiments. *La Riv. Nuovo C* **2022**, *45*, 189–276. [CrossRef]
5. Härkönen, J.; Ott, J.; Gädda, A.; Bezak, M.; Brücken, E.; Tuovinen, E.; Bharthuar, S.; Luukka, P.; Tuominen, E. Processing and interconnections of finely segmented semiconductor pixel detectors for applications in particle physics and photon detection. *Front. Phys.* **2021**, *9*, 601730. [CrossRef]
6. Osadchii, S.M.; Petukhov, A.A.; Dunin, V.B. X-ray Absorption Spectroscopy Analysis of Heavy Metals by Means of a Silicon Detector. *J. Surf. Investig. X-Ray Synchrotron Neutron Tech.* **2019**, *13*, 683–689. [CrossRef]
7. Martin, C.B.; ATLAS Pixel Collaboration. Operational experience and performance with the ATLAS Pixel Detector. *Nucl. Instrum. Methods Phys. Res. Sect. A Accel. Spectrom. Detect. Assoc. Equip.* **2019**, *924*, 293–296. [CrossRef]
8. Hugging, F. The ATLAS pixel detector. *IEEE Trans. Nucl. Sci.* **2006**, *53*, 1732–1736. [CrossRef]
9. Della Negra, M.; Jenni, P.; Virdee, T.S. Journey in the search for the Higgs boson: The ATLAS and CMS experiments at the Large Hadron Collider. *Science* **2012**, *338*, 1560–1568. [CrossRef]
10. Alam, M.S.; Ciocio, A.; Einsweiler, K.; Emes, J.; Gilchriese, M.; Joshi, A.; Kleinfelder, S.; Marchesini, R.; McCormack, F.; Milgrome, O.; et al. The ATLAS silicon pixel sensors. *Nucl. Instrum. Methods Phys. Res. Sect. A Accel. Spectrom. Detect. Assoc. Equip.* **2001**, *456*, 217–232. [CrossRef]
11. Stapnes, S. Detector challenges at the LHC. *Nature* **2007**, *448*, 290–296. [CrossRef] [PubMed]
12. Weber, W.A.; Czernin, J.; Anderson, C.J.; Badawi, R.D.; Barthel, H.; Bengel, F.; Bodei, L.; Buvat, I.; DiCarli, M.; Graham, M.M.; et al. The future of nuclear medicine, molecular imaging, and theranostics. *J. Nucl. Med.* **2020**, *61* (Suppl. S2), 263S–272S. [CrossRef] [PubMed]
13. Enlow, E.; Abbaszadeh, S. State-of-the-art challenges and emerging technologies in radiation detection for nuclear medicine imaging: A review. *Front. Phys.* **2023**, *11*, 1106546. [CrossRef]
14. Lange, J. Recent progress on 3D silicon detectors. *arXiv* **2015**, arXiv:1511.02080. [CrossRef]
15. Parker, S.I.; Kenney, C.J.; Segal, J. 3D—A proposed new architecture for solid-state radiation detectors. *Nucl. Instrum. Methods Phys. Res. Sect. A Accel. Spectrom. Detect. Assoc. Equip.* **1997**, *395*, 328–343. [CrossRef]
16. Kenney, C.; Parker, S.; Segal, J.; Storment, C. Silicon detectors with 3-D electrode arrays: fabrication and initial test results. *IEEE Trans. Nucl. Sci.* **1999**, *46*, 1224–1236. [CrossRef]
17. DaVia, C.; Hasi, J.; Kenney, C.; Kok, A.; Parker, S. 3D silicon detectors—Status and applications. *Nucl. Instrum. Methods Phys. Res. Sect. A Accel. Spectrom. Detect. Assoc. Equip.* **2005**, *549*, 122–125. [CrossRef]
18. Li, Z. New detectors with novel electrode configurations for applications in sLHC and photon sciences. In Proceedings of the Fifteenth Workshop of CERN RD50, Geneva, Switzerland, 16–18 November 2009; CERN: Geneva, Switzerland, 2009.
19. Kenney, C.; Segal, J.; Westbrook, E.; Parker, S.; Hasi, J.; Da Via, C.; Watts, S.; Morse, J. Active-edge planar radiation sensors. *Nucl. Instrum. Methods Phys. Res. Sect. A Accel. Spectrom. Detect. Assoc. Equip.* **2006**, *565*, 272–277. [CrossRef]
20. Li, Z. New BNL 3D-Trench electrode Si detectors for radiation hard detectors for sLHC and for X-ray applications. *Nucl. Instrum. Methods Phys. Res. Sect. A Accel. Spectrom. Detect. Assoc. Equip.* **2011**, *658*, 90–97. [CrossRef]
21. Kalliopuska, J.; Eränen, S.; Virolainen, T. Alternative fabrication process for edgeless detectors on 6 in. wafers. *Nucl. Instrum. Methods Phys. Res. Sect. A Accel. Spectrom. Detect. Assoc. Equip.* **2011**, *633*, S50–S54. [CrossRef]
22. Kalliopuska, J.; Tlustos, L.; Eränen, S.; Virolainen, T. Characterization of edgeless pixel detectors coupled to Medipix2 readout chip. *Nucl. Instrum. Methods Phys. Res. Sect. A Accel. Spectrom. Detect. Assoc. Equip.* **2011**, *648*, S32–S36. [CrossRef]
23. Wu, X.; Kalliopuska, J.; Eränen, S.; Virolainen, T. Recent advances in processing and characterization of edgeless detectors. *J. Instrum.* **2012**, *7*, C02001. [CrossRef]
24. Forcolin, G.; Boscardin, M.; Ficorella, F.; Lai, A.; Loi, A.; Mendicino, R.; Ronchin, S.; Betta, G.-F.D. 3D trenched-electrode pixel sensors: Design, technology and initial results. *Nucl. Instrum. Methods Phys. Res. Sect. A Accel. Spectrom. Detect. Assoc. Equip.* **2020**, *981*, 164437. [CrossRef]
25. Ye, J.; Loi, A.; Lai, A.; Betta, G.-F.D. Design and simulation of modified 3D-trench electrode sensors. *J. Instrum.* **2023**, *18*, C11021. [CrossRef]

26. Liu, M.; Li, X.; Cheng, W.; Li, Z.; Li, Z. Radiation hardness property of ultra-fast 3D-Trench Electrode silicon detector on N-type substrate. *Micromachines* **2021**, *12*, 1400. [CrossRef]
27. Zhu, X.; Li, Z.; Liu, Z.; Long, T.; Zhao, J.; Li, X.; Liu, M.; Wang, M. A Novel Polysilicon-Fill-Strengthened Etch-Through 3D Trench Electrode Detector: Fabrication Methods and Electrical Property Simulations. *Micromachines* **2025**, *16*, 912. [CrossRef]
28. Liu, M.; Lu, S.; Li, Z. Theoretical bases of hypothetical sphere-electrode detectors and practical near-sphere-electrode (semisphere-electrode and near-semisphere-electrode) detectors. *J. Phys. D Appl. Phys.* **2020**, *54*, 045101. [CrossRef]
29. Li, Z. Radiation damage effects in Si materials and detectors and rad-hard Si detectors for SLHC. *J. Instrum.* **2009**, *4*, P03011. [CrossRef]
30. Lindström, G.; Ahmed, M.; Albergo, S.; Allport, P.; Anderson, D.; Andricek, L.; Angarano, M.; Augelli, V.; Bacchetta, N.; Bartalini, P.; et al. Radiation hard silicon detectors—Developments by the RD48 (ROSE) collaboration. *Nucl. Instrum. Methods Phys. Res. Sect. A Accel. Spectrom. Detect. Assoc. Equip.* **2001**, *466*, 308–326. [CrossRef]

micromachines

MDPI

Article

Design of Long-Wave Fully Polarized HgCdTe Photodetector Based on Silicon Metasurface

Bo Cheng [1,2,†], Xiaoming Wang [1,2,†], Yuxiao Zou [3], Guofeng Song [4], Kunpeng Zhai [5,*] and Xiaojun Wang [1,*]

1 Postdoctoral Innovation Practice Base, Chengdu Polytechnic, 83 Tianyi Street, Chengdu 610041, China; chengbo9610@semi.ac.cn (B.C.); ebianwxm@gmail.com (X.W.)
2 Sichuan Provincial Engineering Research Center of Thermoelectric Materials and Devices, Chengdu 610041, China
3 Kunming Institute of Physics, Kunming 650223, China; 13307110322@fudan.edu.cn
4 Institute of Semiconductors, Chinese Academy of Sciences, Beijing 100083, China; sgf@semi.ac.cn
5 Institute of Intelligent Photonics, Nankai University, Tianjin 300071, China
* Correspondence: kpzhai@nankai.edu.cn (K.Z.); 13808193288@163.com (X.W.)
† These authors contributed equally to this work.

Abstract: Polarization-sensitive photodetection is critical for advanced optical systems, yet achieving simultaneous high-fidelity recognition of the circularly polarized (CP) and linearly polarized (LP) light with compact designs remains challenging. Here, we use COMSOL 5.6 software to demonstrate a silicon metasurface-integrated MCT photodetector that resolves both CP and LP signals through a single ultrathin platform. The device deciphers LP states via four orientation-specific linear gratings for differential detection, while chiral symmetric silicon nanostructures enable direct CP discrimination with an exceptional extinction ratio of 30 dB. The proposed architecture combines two breakthroughs: (1) superior polarization reconstruction capability, achieved via the synergy of grating-induced polarization selectivity and chiral near-field enhancement, and (2) a fabrication-simplified process that eliminates multilayer stacking or complex alignment steps. This work establishes a new paradigm for miniaturized, high-performance polarization optics, with potential applications in polarization imaging, quantum communication, and hyperspectral sensing.

Keywords: metasurface; polarization-sensitive photodetection; extinction ratio

1. Introduction

Long-wave infrared photodetectors based on mercury cadmium telluride (MCT) [1] are crucial for applications such as molecular spectroscopy, thermal imaging, and medical diagnosis [2,3]. However, traditional MCT detectors themselves cannot distinguish the polarization state of photons. In scenarios such as chiral molecule detection [4] or polarization thermal radiation analysis [5], the polarization state carries indispensable information. Although linear polarizers [6–8] (such as linear grating structures) can be combined with MCT detectors to extract linear polarization states, research on how to obtain complete Stokes parameters, including circular polarization (CP), in a single chip device is still scarce.

Recent advancements in chiral metasurfaces [9–13] have made it possible to directly detect circularly polarized light through selective transmission or near-field electric field enhancement under left or right circularly polarized (LCP/RCP) illumination. However, most designs only operate in the visible [14–20] or near-infrared bands [21–29], and the designs in the longer wavelength region are relatively scarce [30–33]. This is mainly due to challenges in material loss and manufacturing scalability [34–36]. Additionally, existing long-wave infrared circular polarization-sensitive detectors typically require complex heterostructures or huge external optical components, increasing system complexity and limiting integration potential.

In this work, we designed a chiral MCT detector capable of recognizing circularly polarized light. Its fabrication process is quite simple, requiring only the growth of a silicon thin film followed by one photolithography. We also investigated the electric field distribution of the silicon metasurface and discovered remarkable local electric field enhancement phenomena, which correspond to the transmission minima of RCP and the transmission maxima of LCP. Additionally, we analyzed the influence of the linear grating on the MCT and the potential performance degradation of the micro-device due to process errors.

2. Materials and Methods

The Stokes parameters (S_1, S_2, S_3, S_4) can fully describe the polarization information of light. The first three elements are closely related to linear polarization information, while the fourth element represents circular polarization signals. According to the view of reference [37], to achieve the complete collection of Stokes parameters, six independent polarization pixels are required, which respectively only absorb the 0-degree linearly polarized light signal I_0, 90-degree linearly polarized light signal I_{90}, 45-degree linearly polarized light signal I_{45}, 135-degree linearly polarized light signal I_{135}, left circularly polarized light signal I_{LCP}, and right circularly polarized light signal I_{RCP} in spatial light. The Stokes parameters can be restored through the following formula:

$$S_0 = I_0 + I_{90} \tag{1}$$

$$S_1 = I_0 - I_{90} \tag{2}$$

$$S_2 = I_{45} - I_{135} \tag{3}$$

$$S_3 = I_{LCP} - I_{RCP} \tag{4}$$

When using the technical solution of metasurfaces combined with photodetectors to achieve polarization detection, six metasurfaces with polarization filtering functions are required. Their transmission rates for specific polarized light should be close to zero, while they have considerable transmission rates for the orthogonal polarized light. Generally, there are two performance indicators for the filter plates in polarization detectors. The first one is the circular dichroism (CD), which is the difference in transmission rates of these two orthogonal modes. The higher it is, the more incident light can be converted into electron–hole pairs (or photocurrent). The second one is the extinction ratio (RE), which is the ratio of the transmission rates of these two orthogonal modes or the logarithm of the ratio. The larger it is, the stronger the theoretical ability of Formulas (1)–(4) to restore the polarization signal.

Figure 1a shows the three-dimensional structure of the fully polarized MCT detector, which is composed of a 0-degree linear polarization detector, a 90-degree linear polarization detector, a 45-degree linear polarization detector, a 135-degree linear polarization detector, and left and right circular polarization detectors. Their fabrication process is very simple, requiring only the growth of 3.65 μm of amorphous silicon on the anti-reflection film (MgF2) of a conventional MCT, followed by one photolithography process. The structure of the gratings in the four linear polarization detectors is the same, but their long axes are oriented differently. The silicon metasurfaces in the two circular polarization detectors are mirror-symmetric about the yz plane. Figure 1b is a front view of the 0-degree linear polarization detector, which provides information on the material composition and thickness of the polarization detector. Figure 1c shows the xy cross-section of the linear grating in the 0-degree linear polarization detector, which contains the structure information of the silicon grating. Figure 1d shows the xy cross-section of the unit cell of the chiral metasurface in the left circular polarization detector, which is composed of a rectangle at the center and two quadrilaterals above and below. The upper quadrilateral can be obtained by rotating the lower quadrilateral 180 degrees around the center position. The optical refractive index of the CdZnTe substrate is 2.68, and the refractive index parameters of MCT are from reference [38]. The refractive indices of MgF2 and silicon are 1.38 and 3.48, respectively. The absorption rate of the polarization detector is calculated based on COMSOL software. Periodic boundary conditions are used on the four sides of the optical model to simulate an infinite periodic structure and ensure the continuity of the field between adjacent units. Perfectly matched layer (PML) conditions are used on the upper and lower interfaces of the optical model to absorb outgoing waves and simulate an open boundary with no reflection truncation. The incident port is approximately one wavelength away from the metasurface and can be used to set the initial phase, direction, and polarization of the incident light. The reflectance and transmittance can be obtained by integrating the Poynting vector on the upper and lower boundaries of the model, and the optical absorption rate of the polarization detector is 1 minus the reflectance minus the transmittance.

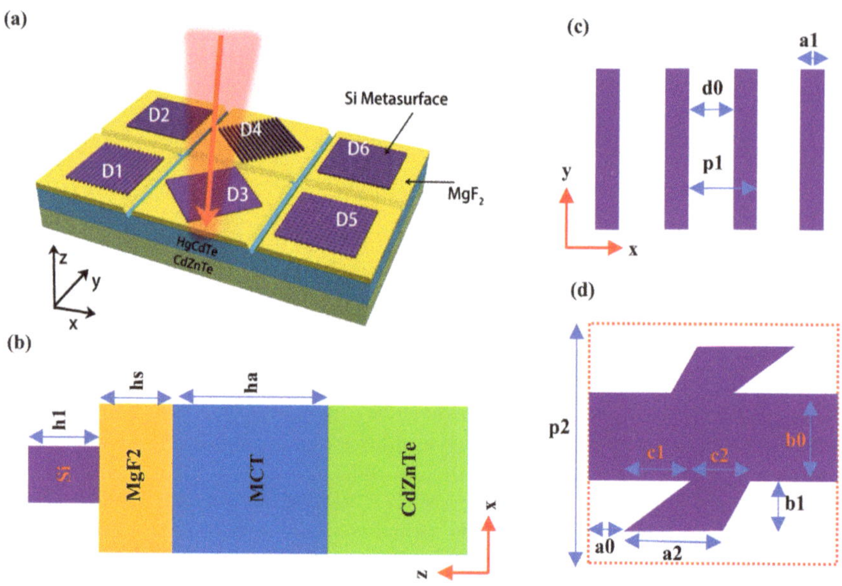

Figure 1. Fully polarized MCT detector. (**a**): Three-dimensional structure diagram of circularly polarized detector and linearly polarized detector. D1, D2, D3, D4, D5, D6 can, respectively, respond to 90-degree linearly polarized light, 0-degree linearly polarized light, 45-degree linearly polarized light, 135-degree linearly polarized light, as well as left circularly polarized light and right circularly polarized light. The six metasurfaces correspond to six independent detector pixels. The polarization detector is isolated through the mesa junction technology to ensure physical separation between the pixels. (**b**) Front view of linearly polarized detector (D1). ha = 10 μm, hs = 4.4 μm, h1 = 3.65 μm. (**c**) xy cross-sectional view of the grating in the linearly polarized detector. p1 = 5.56 μm, d0 = 3.9 μm. (**d**) xy cross-sectional view of the silicon metasurface in the circularly polarized detector. p2 = 5.2 μm, a0 = 0.55 μm, a2 = 1.77 μm, b0 = 1.79 μm, b1 = 1.13 μm, c1 = 0.32 μm, c2 = 2.2 μm.

3. Results

The core components of the six-in-one full polarization MCT detector are the 90-degree linearly polarized MCT and left circularly polarized MCT. The other polarization pixels can be obtained by performing geometric operations on them. Figure 2a shows the absorption of the left circularly polarized MCT for RCP and LCP. It can be found that for RCP, the absorption increases first, then decreases, and then increases again in the 9.4 to 9.8 μm band, with the minimum absorption occurring at a wavelength of 9.6 μm. The maximum absorption peak of LCP also appears near 9.6 μm, with an absorption peak value close to 0.7. The circular polarization extinction ratio (CPER) of the left circularly polarized MCT is $10 \times \log(A_{LCP}/A_{RCP})$, where A_{LCP} and A_{RCP} are the absorption of the polarization pixels for LCP and RCP, respectively. The CPER spectrum is a typical sharp narrow peak, with a peak wavelength of 9.6 μm and a peak intensity of approximately 19 dB. Figure 2b shows the absorption spectrum and linear polarization extinction ratio (LPER) spectrum of the 90-degree linearly polarized MCT. LPER = $10 \times \log(A_{TE}/A_{TM})$, where A_{TE} and A_{TM} are the absorption of the polarization detector for TE and TM, respectively. TE and TM correspond to the cases where the electric field amplitude of the linearly polarized light is in the y and x directions, respectively. For the TM mode, the absorption is close to zero throughout the 9.4 to 9.8 μm band, showing a strong broadband effect. For the TE mode, the absorption increases with the increase in wavelength, with an average absorption rate greater than 0.6. The maximum LPER value occurs at a wavelength of 9.68 μm, approximately 25.5 dB. The LPER at 9.6 μm wavelength is approximately 24 dB.

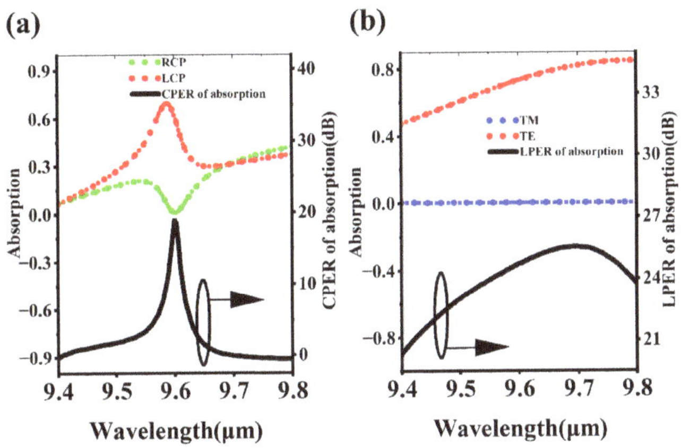

Figure 2. Spectra of polarization detectors. (**a**) Absorption spectrum and CPER spectrum of circular polarization detector. (**b**) Absorption spectrum and LPER spectrum of linear polarization detector. The curves marked with arrows should correspond to the right Y-axis, not the left one.

3.1. *The Polarization Discrimination Capability of the Linear Grating*

It should be emphasized that the MCT detector itself cannot distinguish polarized signals; it requires the assistance of the polarizing grating on its top. For the 90° linearly polarized MCT, the TE mode can penetrate the linear grating and enter the light absorption area of the MCT to form the photoelectric current, while for the TM mode, the linear grating exhibits the function of an optical mirror, preventing photons from passing through it to reach the light absorption area of the MCT. Figure 3a–c show the influence of the grating's structural parameters on the transmission spectrum. As shown in Figure 3a, with the increase in the period *p1*, the transmission rate at a wavelength of 9.6 μm changes very little. Especially for the TM mode, when the wavelength is greater than 10 μm, the absorption spectra are almost identical. Figure 3b indicates the transmission spectrum redshift as the grating width *a1* increases. If the grating width *a1* is increased by just 0.4 µm, the transmission peak of the TE mode can be shifted from 9 µm to 10.5 µm. Figure 3c shows that the transmission spectra of both TE and TM modes exhibit redshift as the grating thickness *h1* increases, but the intensity of the spectral line shift of the TE mode is far more intense than that of the TM mode. Figure 3d presents the transmission spectrum and the linear polarization extinction ratio (LPER) spectrum of the optimized grating. Its maximum CD occurs at a wavelength of approximately 9.8 µm, with a peak close to 1. However, the LPER peak is at a wavelength of 9.6 µm, with a peak value of about 50 dB. Here, LPER represents the grating's extreme suppression effect on the TM mode, the LPER is $10 \times \log(T_{TE}/T_{TM})$, where T_{TE} and T_{TM} are the transmission of the linear grating for TE and TM, respectively. Figure 3e,f show the near-field electric field intensity distribution of the grating. For the TM mode at a wavelength of 9.6 µm, a large red sphere appears inside the silicon grating, which satisfies the typical characteristics of a standing wave. This standing wave exhibits a typical photonic bandgap effect, with almost no free photons being able to radiate into the medium below. For the TE mode at a wavelength of 9.6 µm, the electric field localization sphere in the silicon grating shows a radiative state mode, and it can be found that some photons can leak laterally into the air domain on the side. This radiative state allows the photon energy to penetrate the silicon grating and enter the MgF_2 below.

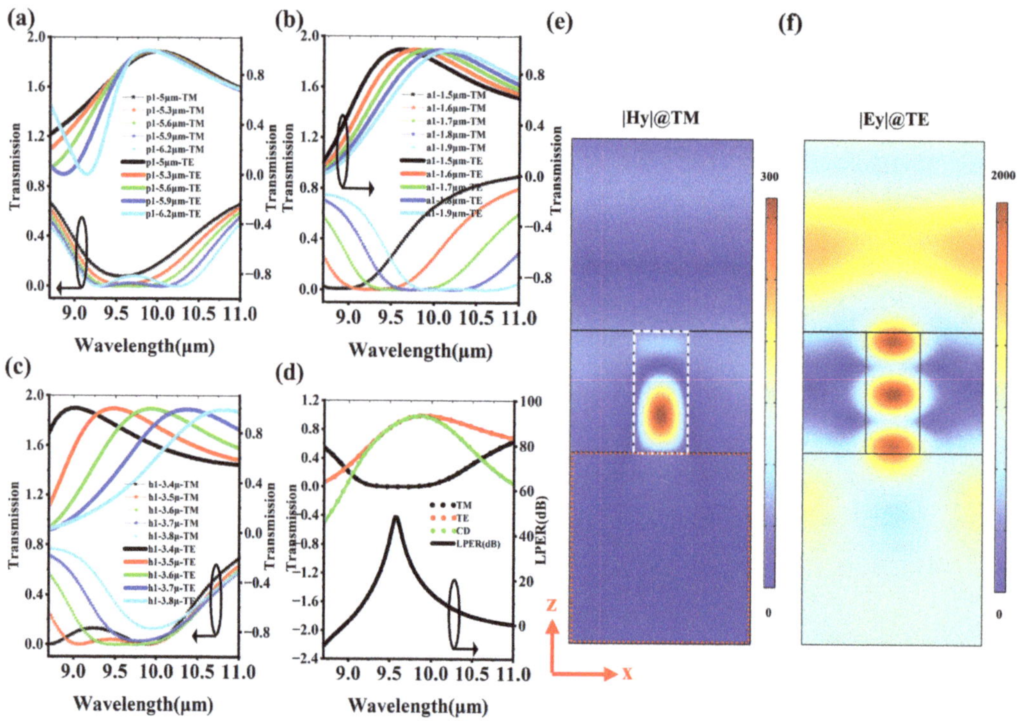

Figure 3. Transmission spectra of the linear grating. (**a**) The influence of the grating period p1. (**b**) The influence of the grating width a1. (**c**) The influence of the grating height h1. (**d**) Transmission spectra and LPER spectra of the line grating. (**e**) Magnetic field distribution in the TM mode. (**f**) Electric field distribution in the TE mode. The white dashed box marks the boundary of the silicon pillar.

3.2. The Circular Dichroism Analysis of Chiral Metasurfaces

Figure 4a,b, respectively, show the amplitude and phase of the elements of the Jones matrix of the chiral metasurface in left circularly polarized MCT, which can describe the transmission characteristics of the chiral metasurface. It can be found that $|Txx| = |Txy| = a$, $|Tyx| = |Tyy| = b$, arg(Tyx − Txx) is approximately 90°, arg(Txy − Txx) is approximately 90°, and arg(Tyy − Txx) is approximately 180°. Therefore, the Jones matrix of the chiral metasurface can be simplified as $T = \begin{bmatrix} a & a*i \\ b*i & -b \end{bmatrix}$. Since $T \times \frac{\sqrt{2}}{2}\begin{bmatrix} 1 \\ i \end{bmatrix} = 0 \times \frac{\sqrt{2}}{2}\begin{bmatrix} 1 \\ i \end{bmatrix}$, $T \times \frac{\sqrt{2}}{2}\begin{bmatrix} 1 \\ -i \end{bmatrix} = 2 \times \begin{bmatrix} a \\ b*i \end{bmatrix}$, where $\frac{\sqrt{2}}{2}\begin{bmatrix} 1 \\ i \end{bmatrix}$ and $\frac{\sqrt{2}}{2}\begin{bmatrix} 1 \\ -i \end{bmatrix}$ correspond to the Jones vectors of RCP and LCP, respectively. The eigenvalue of 0 in the RCP case explains why the transmission rate of RCP is close to 0. Figure 4c shows the electric field localization factor spectrum (EFLF), transmission spectrum, and CPER spectrum of the chiral metasurface. EFLF can represent the strength of the electric field localization, which $\int|E|dV \,/\, \int|E_0|dV$, where E_0 is the electric field intensity of the incident light and E is the electric field inside the silicon metasurface. The integration region is the volume of the silicon metasurface. It can be found that the EFLF of RCP and LCP both have a maximum value near 9.6 μm. Within the wavelength range of 9.4 to 9.8 μm, the maximum transmission of LCP is close to 0.9, the minimum transmission of RCP is close to 0, and the maximum CPER is close to 30 dB. Here, CPER is mainly used to measure the difference in the transmission performance of CP, and it is defined as T_{LCP}/T_{RCP}, where T_{LCP} and T_{RCP} are the transmissions of the metasurface for TE and TM, respectively. Figure 4d–i show the electric field distribution diagrams of the cross-section of the metasurface. It can be found that there are strong electric field localization effects at different cross-sections under different CP modes, which often correspond to the electromagnetic resonances supported by the nanostructures. We boldly speculate that in the RCP incident mode, the incident field and the scattering field formed by the metasurface acting as electromagnetic radiation sources form a phenomenon of destructive interference, cancelling each other out at the transmission end, resulting in a minimum transmission rate. In the LCP incident mode, the incident light and the transmitted electromagnetic field excited by the metasurface form a coherent constructive interference effect, contributing to the formation of the LCP transmission peak.

3.3. The Parameter Optimization of the Spacer Layer *hs*

In the previous section, we emphasized that the polarization discrimination ability stems from the effective screening of different polarized lights by the metasurface, rather than the anisotropic absorption of MCT itself. In addition, to transfer the polarization discrimination ability of the metasurface entirely to the light absorption region of MCT, it is necessary to make the metasurface and MCT two independent optical components as much as possible, which helps to avoid the performance degradation caused by the optical coupling between them. Generally, in the optical field, we can optimize the coupling effect by adjusting the distance between two optical components. Figure 5a shows the influence of the distance *hs* between the metasurface and MCT on the polarization discrimination ability of the linearly polarized detector. The absorption of TM decreases with the increase in *hs*. When *hs* is greater than 3 μm, the absorption remains at a very low and stable level. The absorption of TE shows a cosine function spectrum with the increase in *hs*, which is a regular periodic vibration. In addition, LPER gradually increases with the increase in *hs*. Figure 5b shows the influence of the distance *hs* on the circularly polarized MCT, and its rule is very similar to that in Figure 5a. It is necessary to note an unavoidable phenomenon that the extinction ratio and CD do not increase simultaneously with the increase in *hs*.

Therefore, a new comprehensive performance index for measuring the performance of the polarization detector, the polarization quality factor (PQF), is proposed. It is equal to the sum of the linear polarization quality factor (LPQF) and the circular polarization quality factor (CPQF), where LPQF = CD × LPER and CPQF = CD × CPER, which are used to measure the polarization selection ability of the linearly polarized MCT and the circularly polarized MCT, respectively. When *hs* is less than 3 μm, PQF, LPQF, and CPQF gradually increase with the increase in *hs*. When *hs* is greater than 3 μm, LPQF first increases and then decreases, CPQF first slightly decreases and then continuously increases, and PQF first increases, then remains unchanged, and then continues to increase. Obviously, when *hs* is equal to 6 μm, all three indicators are optimal values. However, considering the difficulty of growing a 6 μm thick MgF2 film, ultimately, *hs* = 4.4 μm is selected as the most appropriate value, as it corresponds to a relatively high PQF value and also represents a manageable difficulty in thin-film growth.

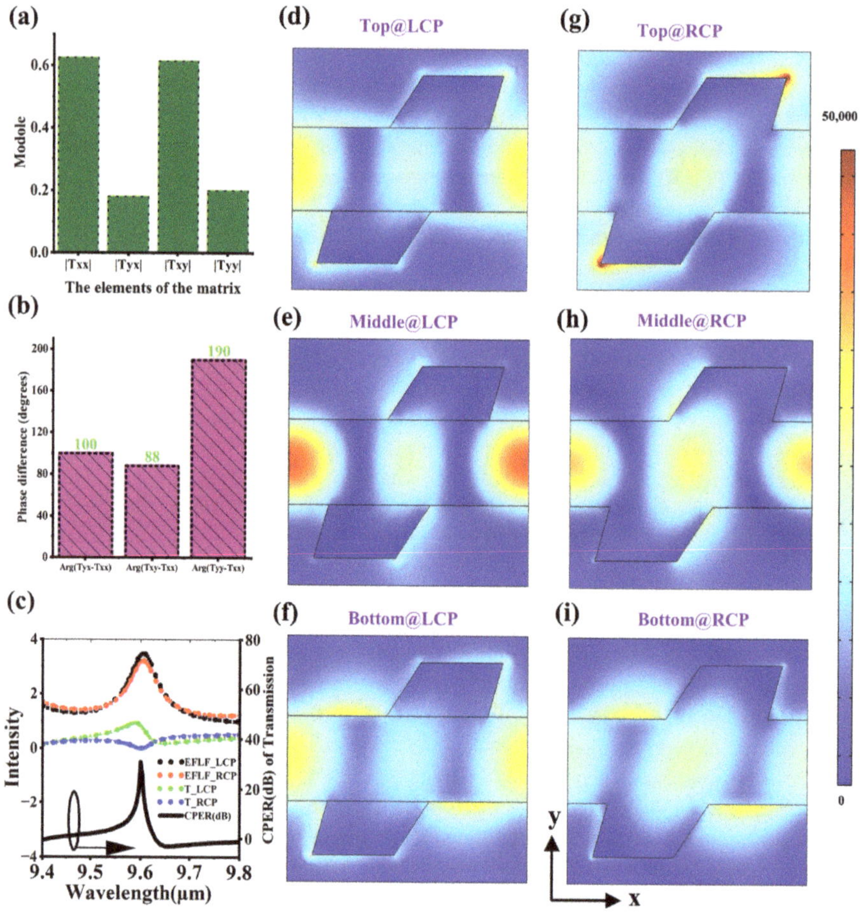

Figure 4. Optical mode analysis of chiral metasurfaces. (**a**) The modulus of the elements of the Jones matrix corresponding to the chiral metasurface. (**b**) The phase difference between the elements of the Jones matrix. (**c**) The EFLF spectrum, transmission spectrum, and CPER spectrum of the chiral metasurface. (**d**–**i**) Electric field distribution maps of the chiral metasurface in CP mode. "Top", "Middle", and "Bottom", respectively, indicate that the cross-section is located at the top, middle, and bottom of the metasurface.

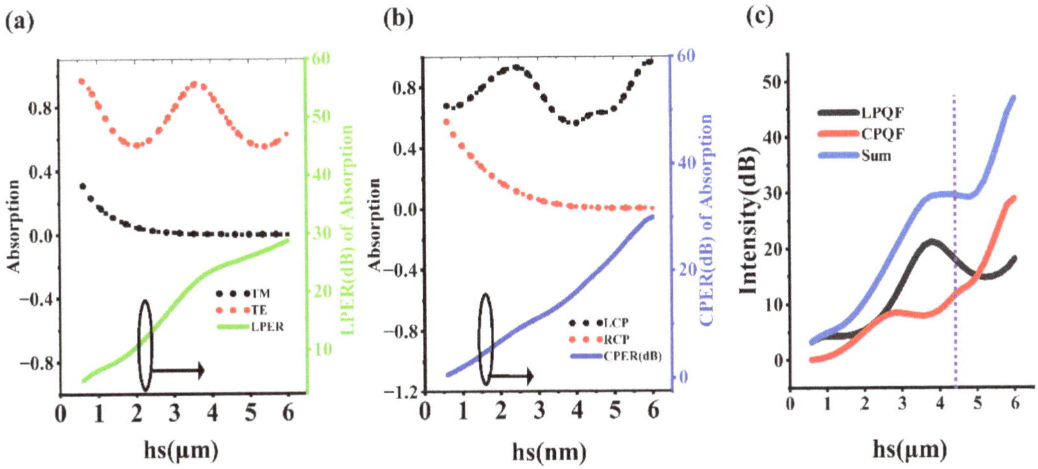

Figure 5. The influence of the thickness *hs* between the metasurface and MCT. (**a**) The influence of *hs* on the absorption and LPER of the linearly polarized MCT. (**b**) The influence of *hs* on the absorption and CPER of the circularly polarized MCT. (**c**) LPQF and CPQF related to *hs*.

3.4. The Analysis of Potential Errors

In the actual fabrication process of metasurfaces, many process errors can cause the structure of the metasurface to deviate from the ideal value, ultimately resulting in performance loss. After photolithography and etching, the rectangular structure usually turns into a trapezoid due to the difficulty in ensuring 100% anisotropy in the process, as shown in the inset of Figure 6a. Figure 6a shows the influence of the etching tilt angle (i.e., the lower base angle of the trapezoid) on the performance of the linearly polarized MCT. It can be found that as the etching tilt angle decreases, the LPQF also decreases slowly. When the etching tilt angle is 80 degrees, the LPQF is reduced by half. Figure 6b shows the under-etching and over-etching situations of the silicon metasurface, indicating the etching residue of the silicon material. Figure 6c shows that when the under-etching depth error increases from 0 nm to 200 nm, the LPQF decreases from 18 dB to 14 dB. Figure 6d shows that the over-etching depth error has almost no effect on the LPQF value. Figure 6e shows a two-dimensional optical simulation model for calculating the number of unit cells of the linear grating, with perfect matching layer conditions applied to the upper and lower end faces and scattering boundary conditions (SBC) applied to the left and right end faces. As shown in Figure 6f, as the number of unit cells increases, the dichroism of the linearly polarized detector first increases and then remains unchanged, with the inflection point corresponding to approximately six unit cells.

Figure 7 illustrates the influence of possible errors in the circularly polarized MCT. Figure 7a shows the impact of the grid size of the absorption region in the MCT and the grid size of the silicon metasurface. The variation in the MCT grid size within the range of 30 nanometers to 400 nanometers does not have a significant effect on the absorption of circularly polarized light. When the grid size of the silicon metasurface exceeds 0.5 μm, the absorption of LCP drops sharply, while that of RCP increases rapidly. Additionally, due to process factors such as photoresist reflow, insufficient anisotropy during etching, ion bombardment, or thermal diffusion effects, edge smoothing occurs, causing rectangular structures to become rounded (e.g., the edges of grooves or patterns become rounded); this special rounded corner phenomenon cannot be ignored either. Figure 7b shows the degree of damage to the performance of the circularly polarized detector caused by the radius of the etched round corners. When the radius of the etched round corners increases from

0 nm to 300 nm, the CPQF decreases from 11.5 dB to 9 dB. As shown in Figure 7c, with the increase in the under-etching depth error, the absorption of LCP first increases and then decreases, while that of RCP keeps increasing. When the etching depth error is 140 nm, the circular dichroism has already become very low. Figure 7d demonstrates the impact of over-etching. Although the error in the over-etching depth also has an adverse effect on the circularly polarized detector, its degree of damage is far less than that of under-etching. Given this situation, when etching the silicon metasurface, the etching time can be increased by a few seconds to ensure that the etching is in a critical etching or over-etching state. In addition, the influence of the incident angle and the MgF_2 film on the micro-device can be seen in the Supplementary Materials.

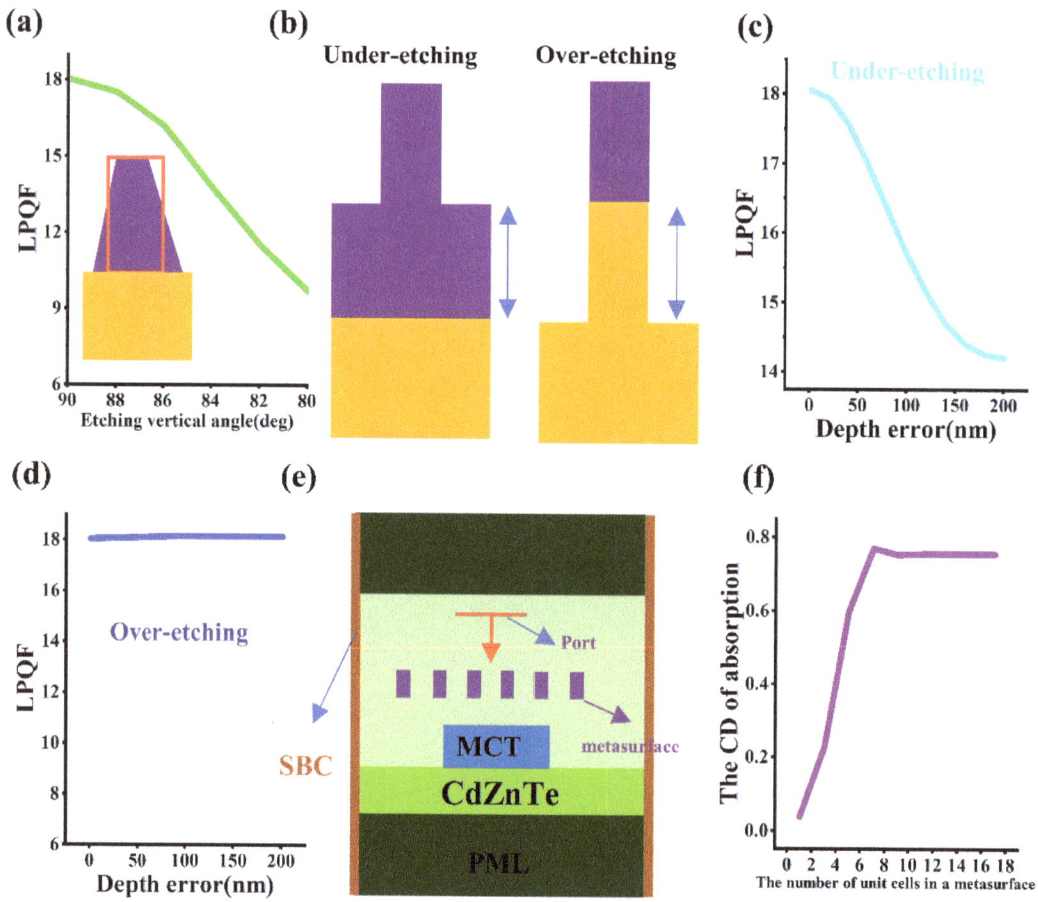

Figure 6. Error analysis of the linearly polarized MCT. (**a**) Influence of etching tilt angle. (**b**) Geometric diagrams of the under-etching and over-etching cases. (**c**) Influence of the under-etching depth error. (**d**) Influence of the over-etching depth error. (**e**) Schematic diagram of a two-dimensional simulation model for analyzing the number of unit cells of metasurface. (**f**) Influence of the number of unit cells of metasurface.

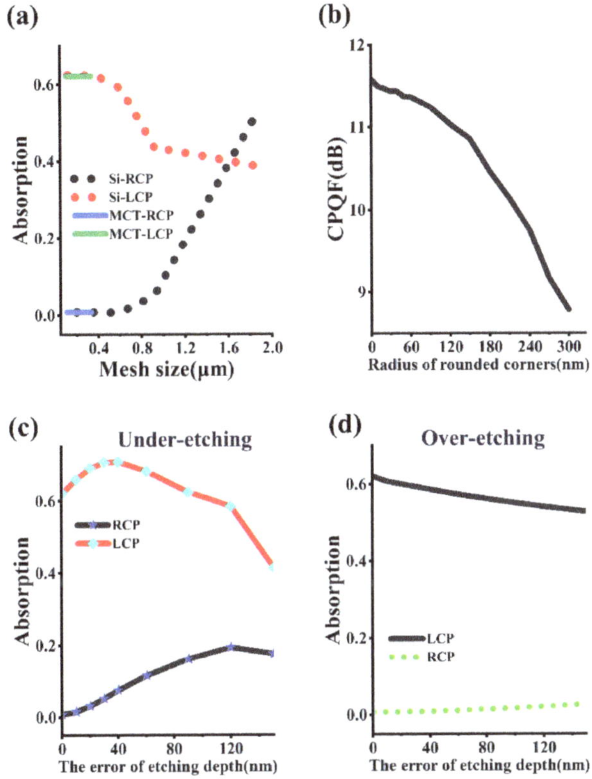

Figure 7. Error analysis of circularly polarized MCT. (**a**) The influence of the grid size. (**b**) The influence of the filet radius. (**c**) The influence of under-etching depth error. (**d**) The influence of over-etching depth error.

4. Discussion

In summary, we demonstrate a monolithic MCT photodetector with circular polarization resolution capability at a wavelength of 9.6 μm, which is achieved by directly integrating a Si metasurface with ultra-large chiral transmission difference. The detector has a huge polarization extinction ratio of 19 dB for LCP and RCP. Crucially, when combined with a linearly polarized MCT based on a wire grating, our six-in-one device forms a compact photonic system capable of enabling full polarization reconstruction without the need for bulky optical components. This design bridges the gap between polarization optics and semiconductor photodetection, providing a scalable solution for miniaturized long-wave infrared polarization imaging and sensing applications.

Supplementary Materials: The following supporting information can be downloaded at: https://www.mdpi.com/article/10.3390/mi16080937/s1, Figure S1. The influence of the incident angle. (a) The incident plane is the XZ plane. (b) The incident plane is the YZ plane. Figure S2. The influence of the changes in thickness and refractive index of the spacer layer caused by the temperature effect. (a) The variation in the thickness of the spacer layer Δhs. (b) The variation in the refractive index of the spacer layer Δn. Table S1: Meshing strategy of the model.

Author Contributions: Conceptualization, B.C. and Y.Z.; methodology, B.C.; software, B.C.; validation, B.C., G.S. and X.W. (Xiaoming Wang); formal analysis, K.Z., X.W. (Xiaojun Wang); investigation, K.Z., X.W. (Xiaojun Wang); resources, Y.Z.; data curation, B.C.; writing—original draft preparation,

B.C.; writing—review and editing, K.Z., X.W. (Xiaoming Wang) and X.W. (Xiaojun Wang); visualization, Y.Z.; supervision, Y.Z.; project administration, Y.Z.; funding acquisition, K.Z. All authors have read and agreed to the published version of the manuscript.

Funding: This research was funded by Natural Science Foundation of Tianjin, grant number NO. 24JC-QNJC01880; Young Elite Scientists Sponsorship Program of Tianjin, grant number NO. YESS20240775; the Strategic Priority Research Program of the Chinese Academy of Sciences (grant no. XDB0980000).

Data Availability Statement: The raw data supporting the conclusions of this article will be made available by the authors on request.

Acknowledgments: Thanks to Liu Jietao from Nankai University for his support on the software.

Conflicts of Interest: The authors declare no conflicts of interest.

References

1. Rogalski, A. HgCdTe infrared detector material: History, status and outlook. *Rep. Prog. Phys.* **2005**, *68*, 2267–2336. [CrossRef]
2. Eich, D.; Schirmacher, W.; Hanna, S.; Mahlein, K.M.; Fries, P.; Figgemeier, H. Progress of MCT Detector Technology at AIM Towards Smaller Pitch and Lower Dark Current. *J. Electron. Mater.* **2017**, *46*, 5448–5457. [CrossRef]
3. Marcott, C.; Reeder, R.C. FT-IR spectroscopic imaging microscopy using an MCT focal-plane array detector. *Am. Inst. Phys.* **1998**, *430*, 377–378.
4. Anet, F.A.L.; Kopelevich, M. Detection and assignments of diastereotopic chemical shifts in partially deuteriated methyl groups of a chiral molecule. *Cheminform* **1989**, *111*, 3429–3431.
5. Troitsky, A.V.; Osharin, A.M.; Korolev, A.V.; Strapp, J.W. Polarization of Thermal Microwave Atmospheric Radiation Due to Scattering by Ice Particles in Clouds. *J. Atmos. Sci.* **2003**, *60*, 1608–1620. [CrossRef]
6. Deng, Y.; Meng, C.; Thrane, P.C.V.; Sande, S.I.; Bozhevolnyi, S.I.; Ding, F. MEMS-integrated metasurfaces for dynamic linear polarizers. *Optica* **2024**, *11*, 7. [CrossRef]
7. Han, C.; Tam, W.Y. Plasmonic ultra-broadband polarizers based on Ag nano wire-slit arrays. *Appl. Phys. Lett.* **2015**, *106*, 3431–3433. [CrossRef]
8. Cui, Y.; Azzam, R.M.A. Sixteen-beam grating-based division-of-amplitude photopolarimeter. *Opt. Lett.* **1996**, *21*, 89–91. [CrossRef]
9. Wu, C.; Arju, N.; Fan, J.; Brener, I.; Shvets, G. Spectrally selective chiral silicon metasurfaces based on infrared Fano resonances. In Proceedings of the Lasers & Electro-optics, San Jose, CA, USA, 8–13 June 2014.
10. Li, Z.; Liu, W.; Cheng, H.; Chen, S.; Tian, J. Tunable dual-band asymmetric transmission for circularly polarized waves with graphene planar chiral metasurfaces. *Opt. Lett.* **2016**, *41*, 3142–3145. [CrossRef]
11. Zhang, N.; Gao, F.; Wang, R.; Shen, Z.; Han, D.; Cui, Y.; Zhang, L.; Chang, C.; Qiu, C.W.; Chen, X. Deep-Learning Empowered Customized Chiral Metasurface for Calibration-Free Biosensing. *Adv. Mater.* **2025**, *37*, e2411490. [CrossRef]
12. Gromyko, D.; Loh, J.S.; Feng, J.; Qiu, C.W.; Wu, L. Enabling All-to-Circular Polarization Up-Conversion by Nonlinear Chiral Metasurfaces with Rotational Symmetry. *Phys. Rev. Lett.* **2025**, *134*, 023804. [CrossRef]
13. Ma, C.; Yu, P.; Jing, Z.; Zhu, Y.; Li, P.; Wang, W.; Xu, H.; Zhang, Y.; Pan, L.; Choi, T.-Y.; et al. Circular polarization-selective optical, photothermal, and optofluidic effects in chiral metasurfaces. *Photonics Res.* **2024**, *12*, 331. [CrossRef]
14. Zhao, Y.; Belkin, M.A.; Alù, A. Twisted optical metamaterials for planarized ultrathin broadband circular polarizers. *Nat. Commun.* **2012**, *3*, 870. [CrossRef]
15. Dietrich, K.; Menzel, C.; Lehr, D.; Puffky, O.; Huebner, U.; Pertsch, T.; Tuennermann, A.; Kley, E.B. Elevating optical activity: Efficient on-edge lithography of three-dimensional starfish metamaterial. *Appl. Phys. Lett.* **2014**, *104*, 2517–2534. [CrossRef]
16. Bachman, K.A.; Peltzer, J.J.; Flammer, P.D.; Furtak, T.E.; Collins, R.T.; Hollingsworth, R.E. Spiral plasmonic nanoantennas as circular polarization transmission filters. *Opt. Express* **2012**, *20*, 1308–1319. [CrossRef] [PubMed]
17. Fedotov, V.A.; Schwanecke, A.S.; Zheludev, N.I.; Khardikov, V.V.; Prosvirnin, S.L. Asymmetric Transmission of Light and Enantiomerically Sensitive Plasmon Resonance in Planar Chiral Nanostructures. *Nano Lett.* **2007**, *7*, 1996–1999. [CrossRef]
18. Cadusch, J.J.; James, T.D.; Djalalian-Assl, A.; Davis, T.J.; Roberts, A. A Chiral Plasmonic Metasurface Circular Polarization Filter. *IEEE Photonics Technol. Lett.* **2014**, *26*, 2357–2360. [CrossRef]
19. Zuo, J.; Bai, J.; Choi, S.; Basiri, A.; Chen, X.; Wang, C.; Yao, Y. Chip-integrated metasurface full-Stokes polarimetric imaging sensor. *Light Sci. Appl.* **2023**, *12*, 218. [CrossRef]
20. Li, L.W.; Rubin, N.A.; Juhl, M.; Park, J.-S.; Capasso, F. Evaluation and characterization of imaging polarimetry through metasurface polarization gratings. *Appl. Opt.* **2023**, *62*, 1704–1722. [CrossRef]
21. Li, W.; Coppens, Z.J.; Besteiro, L.V.; Wang, W.; Govorov, A.O.; Valentine, J. Circularly polarized light detection with hot electrons in chiral plasmonic metamaterials. *Nat. Commun.* **2015**, *6*, 8379. [CrossRef]

22. Maksimov, A.A.; Tartakovskii, I.I.; Filatov, E.V.; Lobanov, S.V.; Gippius, N.A.; Tikhodeev, S.G.; Schneider, C.; Kamp, M.; Maier, S.; Höfling, S.; et al. Circularly polarized light emission from chiral spatially-structured planar semiconductor microcavities. *Phys. Rev. B* **2014**, *89*, 107–114. [CrossRef]
23. Lobanov, S.V.; Weiss, T.; Gippius, N.A.; Tikhodeev, S.G.; Kulakovskii, V.D.; Konishi, K.; Kuwata-Gonokami, M. Polarization control of quantum dot emission by chiral photonic crystal slabs. *Opt. Lett.* **2015**, *40*, 1528. [CrossRef] [PubMed]
24. Lobanov, S.V.; Tikhodeev, S.G.; Gippius, N.A.; Maksimov, A.A.; Hfling, S. Controlling circular polarization of light emitted by quantum dots using chiral photonic crystal slab. *Phys. Rev. B* **2015**, *92*, 205309. [CrossRef]
25. Basiri, A.; Chen, X.; Bai, J.; Amrollahi, P.; Carpenter, J.; Holman, Z.; Wang, C.; Yao, Y. Nature-inspired chiral metasurfaces for circular polarization detection and full-Stokes polarimetric measurements. *Light. Sci. Appl.* **2019**, *8*, 78. [CrossRef]
26. Hu, J.; Zhang, C.; Dong, Y.; Zeng, A.; Wang, C. High efficiency all-dielectric pixelated metasurface for near-infrared full-Stokes polarization detection. *Photonics Res.* **2021**, *9*, 583–589.
27. Cheng, B.; Zou, Y.; Shao, H.; Li, T.; Song, G. Full-Stokes imaging polarimetry based on a metallic metasurface. *Opt. Express* **2020**, *28*, 27324–27336. [CrossRef]
28. Cheng, B.; Zou, Y.; Song, G. Ultra-large bandwidth fully polarized photodetectors based on displacement-induced chiral dielectric metasurfaces. *Appl. Opt.* **2025**, *64*, 2332–2338. [CrossRef]
29. Cheng, B.; Zou, Y.; Song, G. Full-stokes polarization photodetector based on the chiral metasurface with the dislocated double gold rod configurations. *Opt. Laser Technol.* **2024**, *174*, 110531. [CrossRef]
30. Miyazaki, H.T.; Mano, T.; Kasaya, T.; Osato, H.; Watanabe, K.; Sugimoto, Y.; Kawazu, T.; Arai, Y.; Shigetou, A.; Ochiai, T.; et al. Synchronously wired infrared antennas for resonant single-quantum-well photodetection up to room temperature. *Nat. Commun.* **2020**, *11*, 565. [CrossRef]
31. Ogawa, S.; Kimata, M. Wavelength- or Polarization-Selective Thermal Infrared Detectors for Multi-Color or Polarimetric Imaging Using Plasmonics and Metamaterials. *Materials* **2017**, *10*, 493. [CrossRef]
32. Chen, C.; Huang, Y.; Wu, K.; Bifano, T.G.; Anderson, S.W.; Zhao, X.; Zhang, X. Polarization insensitive, metamaterial absorber-enhanced long-wave infrared detector. *Opt. Express* **2020**, *28*, 28843–28857. [CrossRef]
33. Tan, X.; Zhang, H.; Li, J.; Wan, H.; Guo, Q.; Zhu, H.; Liu, H.; Yi, F. Non-dispersive infrared multi-gas sensing via nanoantenna integrated narrowband detectors. *Nat. Commun.* **2020**, *11*, 5245. [CrossRef]
34. Pustelny, T.; Struk, P.; Mergo, P.; Wojtas, J.; Martyniuk, P.; Kopytko, M.; Grodecki, K.; Gawron, W.; Gomuka, E. High operating temperature long-wave HgCdTe detector for fast response operation: Optimization approach. In Proceedings of the 11th Conference on Integrated Optics: Sensors, Sensing Structures, and Methods, Szczyrk, Poland, 29 February–4 March 2016.
35. Forrest, W. Development of Long-Wave HgCdTe Detector Arrays for Future Space Missions. *NASA APRA Propos.* **2012**, 12-APRA12.
36. Hu, W.; Ye, Z.; Liao, L.; Chen, H.; Lu, W. A 128×128 long-wavelength/mid-wavelength two-color HgCdTe infrared focal plane array detector with ultra-low spectral crosstalk. *Opt. Lett.* **2014**, *39*, 5184–5187. [CrossRef]
37. Ehsan, A.; Mahsa, K.S.; Amir, A.; Andrei, F. Full Stokes imaging polarimetry using dielectric metasurfaces. *ACS Photonics* **2018**, *5*, 3132–3140. [CrossRef]
38. Sen, S.; Rhiger, D.R.; Curtis, C.R.; Kalisher, M.H.; Hettich, H.L.; Currie, M.C. Infrared absorption behavior in CdZnTe substrates. *J. Electron. Mater.* **2001**, *30*, 611–618. [CrossRef]

micromachines

MDPI

Article

A Novel Polysilicon-Fill-Strengthened Etch-Through 3D Trench Electrode Detector: Fabrication Methods and Electrical Property Simulations

Xuran Zhu [1,2], Zheng Li [1,2,*], Zhiyu Liu [1,2], Tao Long [1,2], Jun Zhao [1,2], Xinqing Li [2,3], Manwen Liu [4] and Meishan Wang [1,2]

1 School of Integrated Circuits, Ludong University, Yantai 264025, China
2 Engineering Research Center of Photo Detector Special Chip in Universities of Shandong, Ludong University, Yantai 264025, China
3 School of Materials Science and Engineering, Xiangtan University, Xiangtan 411105, China
4 Institute of Microelectronics, Chinese Academy of Sciences, Beijing 100029, China
* Correspondence: 3636@ldu.edu.cn

Abstract: Three-dimensional trench electrode silicon detectors play an important role in particle physics research, nuclear radiation detection, and other fields. A novel polysilicon-fill-strengthened etch-through 3D trench electrode detector is proposed to address the shortcomings of traditional 3D trench electrode silicon detectors; for example, the distribution of non-uniform electric fields, asymmetric electric potential, and dead zone. The physical properties of the detector have been extensively and systematically studied. This study simulated the electric field, potential, electron concentration distribution, complete depletion voltage, leakage current, capacitance, transient current induced by incident particles, and weighting field distribution of the detector. It also systematically studied and analyzed the electrical characteristics of the detector. Compared to traditional 3D trench electrode silicon detectors, this new detector adopts a manufacturing process of double-side etching technology and double-side filling technology, which results in a more sensitive detector volume and higher electric field uniformity. In addition, the size of the detector unit is 120 μm × 120 μm × 340 μm; the structure has a small fully depleted voltage, reaching a fully depleted state at around 1.4 V, with a saturation leakage current of approximately 4.8×10^{-10} A, and a geometric capacitance of about 99 fF.

Keywords: novel polysilicon-fill-strengthened etch-through 3D trench electrode; full 3D simulation; electric field distribution; electric potential distribution; fully depleted voltage; induced transient current

1. Introduction

At present, as society continues to progress and science and technology keep advancing, the application of semiconductor materials has become an indispensable part of life. The detector technology based on semiconductor materials has emerged and matured. Among various types of semiconductor detectors, silicon detectors have many advantages, such as high energy and position resolution, fast response speed, high sensitivity, and ease of integration. They have been widely used in fields such as high-energy physics [1–3], astrophysics, nuclear medicine [4,5], X-ray imaging [6], national defense, industry [7], high-energy physics experiments in high-energy particle detection, and many others. The commonly used thickness of silicon detectors is about 300 μm. With the development of particle physics and related fields, research on silicon detectors will continue to be carried out both domestically and internationally.

The common types of silicon detectors mainly include silicon microstrip detectors [8,9], silicon drift detectors [10,11], three-dimensional columnar electrode silicon detectors [12–14], and three-dimensional trench electrode silicon detectors [15,16], which work in a similar way. With the evolution and development of various types of semiconductor detectors from 2D to 3D, the structure and performance of detectors are constantly being improved and optimized, and 3D silicon detectors have been proven to be more radiation resistant than silicon planar detectors [13]. Although a double-sided detector concept was proposed in Ref. [16] as well, no details were provided on the detector structures and processing technologies. The first non-etch-through 3D trench electrode detector was first fabricated and tested in CNM, with the results reported in Ref. [16].

The trench wall electrode (straight) concept, initially proposed by C. Kenney [17], employs embedded electrode structures within silicon substrates. This technologically demanding approach requires advanced design methodologies and precision manufacturing techniques. In 1997, S.Parker designed and developed a structure completely different from planar detectors, called a three-dimensional columnar electrode silicon detector. In 2005, Brookhaven Laboratory in the United States proposed a new type of 3D electrode detector, the 3D trench electrode silicon detector. The electric field distribution within the detector unit of this trench-shaped electrode is more uniform, and the isolation of each unit is stronger when forming an array. But for traditional 3D trench electrode silicon detectors, it is necessary to leave an un-etched thickness to ensure that the detector body does not detach from the substrate during the etching process. After etching, due to the thin un-etched thickness of the silicon substrate, the trench electrode cannot completely penetrate the entire silicon substrate and usually leaves about 10 percent of the depth of the silicon substrate. The un-etched substrate here is the dead zone, which is a weak or zero-electric-field region that exists in the detector [18]. If the dead zone occupies a large part of the detector volume, the electrical characteristics of the detector will be severely affected, thereby reducing the charge collection efficiency.

In order to address the aforementioned shortcomings of traditional 3D trench electrode silicon detectors, this paper proposes a novel polysilicon-fill-strengthened etch-through 3D trench electrode detector and systematically studies its electrical properties. This new type of detector is etched and filled on both sides of the silicon detector to form a through trench electrode, which is embedded in the silicon body. Its depletion electric field is perpendicular to the thickness direction of the silicon wafer, so its depletion voltage is independent of thickness. Each pixel unit in the detector is isolated from other pixel units by highly doped trenches, making it an electrically independent unit. And adjacent detector units can also share the same trench electrode, which greatly reduces dead zones and electrode connections. In this article, the simulation assistance software was used for electrical simulation and performance comparison of the detector.

2. The Structure of a Novel Polysilicon-Fill-Strengthened Etch-Through 3D Trench Electrode Detector on N-Type Bulk

2.1. The Model of a Novel Polysilicon-Fill-Strengthened Etch-Through 3D Trench Electrode Detector

The most obvious advantage of 3D electrode detectors is its radiation hardness, which has wide applications in high-energy physics experiments. In this work, however, we are focusing our detector in the area of X-ray detection and imaging with little radiation effect. Therefore, we can use a relatively large detector size and thickness in the realm of hundreds of microns. When selecting a square as the shape of the detector unit, the practical advantage is that it can easily generate a detector array, and the resulting pixel detector also has a regular shape that can be easily cut from the chip and placed in the application device. The detector bulk is N-type Si with $1 \times 10^{12}/\text{cm}^3$ phosphorus doping with a thickness of 340 μm. The central collection column electrode is doped with n^+ at a concentration of $1 \times 10^{19}/\text{cm}^3$, and the 3D trench electrode is doped with p^+ at a concentration of $1 \times 10^{19}/\text{cm}^3$. The p-n junction is placed near the 3D trench electrode to ensure good electric field distribution and low full depletion voltage of the detector. At the top of the detector, the center column electrode is covered with a 1 μm layer of aluminum, and the area between the center column electrode and the etched trench electrode is covered with a 1 μm layer of silicon oxide. At the bottom of the detector, the trench electrode is covered with a 1 μm layer of aluminum, and the gap between the trench electrode and the etched columnar region is covered with a 1 μm layer of silicon oxide. The trench and column are filled with polysilicon. Figure 1 shows a schematic diagram of the structure of a novel polysilicon-fill-strengthened etch-through 3D trench electrode detector.

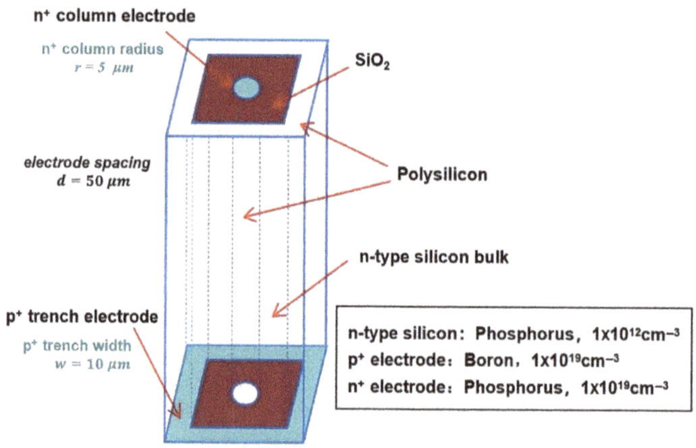

Figure 1. A novel polysilicon-fill-strengthened etch-through 3D trench electrode detector structure.

2.2. The Technological Process of a Novel Polysilicon-Fill-Strengthened Etch-Through 3D Trench Electrode Detector

The primary technical challenge in fabricating 3D electrode detectors is how to form precisely trench-shaped conductive elements. In 2025, Liu, M. and others proposed and simulated a double-sided 3D trench electrode detector structure; in the simulation, the wafer thickness of the detector unit was set to 700 μm, and the width of the column electrode and trench electrode were the same, both 5 μm. The dark current of this structure was about 10^{-10}A, and the capacitance was less than 250 fF. The acquisition time was on the order of tens of nanoseconds. Through Bosch deep reactive ion etching (DRIE), the thickness of the 3D trench electrode can reach 311 μm, and the aspect ratio can reach 105:1 [19], with a precision of double-sided alignment better than 2 μm. Since our trench width is 10 μm, this allows an acceptable alignment of the top and bottom trenches, thereby enabling practical implementation of such advanced devices. Since, during the process of trench filling of polysilicon, the top and bottom surface will be protected, there will be no polysilicon layers formed on both surfaces, and therefore there will be no bowing/warping effects.

Taking a 3 × 3 detector array as an example, we will introduce the manufacturing method of the detector here. The majority charge carrier in N-type silicon is electron, with an electron mobility of approximately 1450 $cm^2/s/V$, significantly higher than the mobility of holes in P-type silicon, which is 450 $cm^2/s/V$. However, for applications in high-radiation environments, it is recommended to use a P-type substrate configuration for its radiation hardness [20]. In this case, the peripheral trenches of the device need to be heavily doped with N^+ (Phosphorus), and the central columnar electrode doped with P^+ (Boron), which becomes a complicated P-type process. Therefore, for our cases of non-irradiation or light irradiation conditions, a lightly doped N-type silicon substrate is preferred for simplicity in detector process. As shown in Figure 2, the first part of the detector fabrication process steps is as follows: (a) select lightly doped N-type silicon, and cover the upper and lower surfaces with 1 μm of SiO_2 (thermally grown SiO_2); (b) etch half of the silicon body to form a trench electrode of a depth of half of the wafer and a width of 5–20 μm, doping the trench walls with P^+ (Boron) by diffusion or ion implantation (with a special angle related to the trench depth and width), with P-type heavy doping at a doping concentration of $1 \times 10^{18}/cm^3$–$2 \times 10^{20}/cm^3$; (c) conduct polysilicon filling, where the polysilicon deposition method can be chemical vapor deposition (CVD); (d) flip over the wafer to the bottom side; and (e) for P^+-trench wall etching, dope the trench walls with P^+ (Boron) by diffusion or ion implantation, with P-type heavy doping at a doping concentration of $1 \times 10^{18}/cm^3$–$2 \times 10^{20}/cm^3$. Figure 3 shows the second part of the detector fabrication process steps as follows: (a) polysilicon filling; (b) N^+-column etching, doping the columns with N^+ (Phosphorus), with N-type heavy doping at a doping concentration of $1 \times 10^{18}/cm^3$–$2 \times 10^{20}/cm^3$; (c) Al-coating on P^+-trench with a thickness of 1 μm (trench electrodes will be isolated by the negative surface charge on the silicon surface induced by positive oxide charge); (d) flip over the wafer to the top side; (e) Al-coating on the top side (N^+-columns are isolated by the surrounding P^+ trenches); and (f) connection to readout electronics. A SiO_2 insulating layer isolates the anode and cathode, exhibiting an initial oxide charge density of $4 \times 10^{11}/cm^2$. This value demonstrates a positive correlation with radiation intensity, potentially escalating under prolonged exposure.

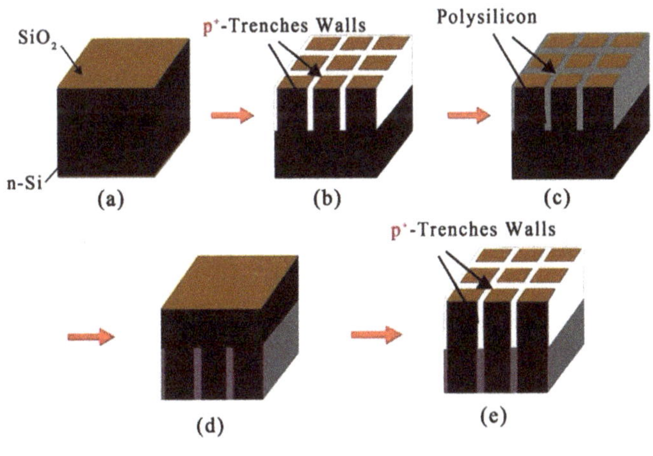

Figure 2. Novel polysilicon-fill-strengthened etch-through 3D trench electrode detector fabrication process steps (part one).

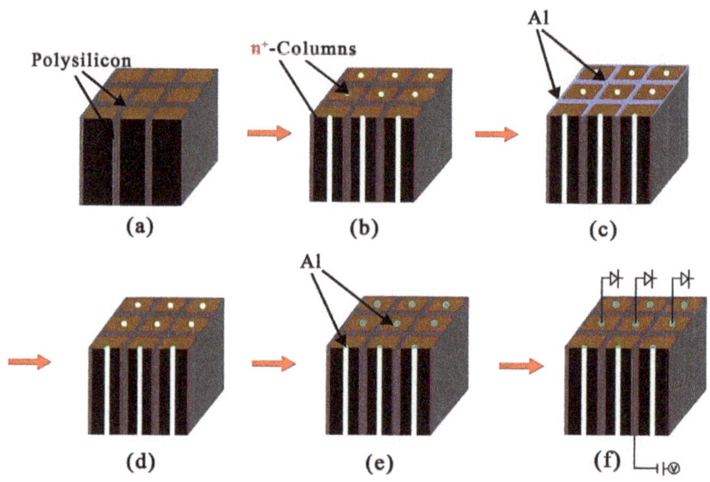

Figure 3. Novel polysilicon-fill-strengthened etch-through 3D trench electrode detector fabrication process steps (part two).

3. Electrical Characteristic Results

3.1. Three-Dimensional Simulation of a Novel Polysilicon-Fill-Strengthened Etch-Through 3D Trench Electrode Detector

This part of the simulation was conducted using the full 3D simulation assistance software, which can accurately simulate complex three-dimensional structures and their electrical properties. The physical models used in our simulations include SRH recombination, doping-dependent carrier mobility (Masetti model), and high field saturation effect. TCAD contains many comprehensive components that users can directly call as needed. Components have Sentaurus Workbench, Sentaurus Device, Sentaurus Process, Sentaurus Structure Editor, Mesh and Noffset 3D, Ligament, Tecplot SV, Inspect, and Calibration Kit Wait. The fixed charge density of the oxide layer was set to be $4 \times 10^{11}/\text{cm}^2$ to simulate the influence of oxide charge. Figures 4a and 4b, respectively, present schematic diagrams of the upper and lower surfaces of the detector unit, visually demonstrating the positions and dimensions of the anode and cathode. In our study, we chose a detector unit cell with the following parameters: cell size 110 µm, trench width 10 µm, central column diameter 10 µm, and detector thickness 340 µm. In this case, the relative inactive volume due to trenches and column is 18%. We note that if we increased cell size, this relative inactive volume was greatly reduced. For example, if we increased cell size to 170 µm, the relative inactive volume will be reduced to 11.7%. Figure 4c provides a cross-sectional view of the structure, with axial units in micrometers (µm). By means of the cross-sectional view, the penetration of the trench electrode and the distribution of material layers can be more intuitively observed, providing important references for subsequent performance analysis and optimization. A negative bias voltage is applied to the cathode electron on the bottom side, and the anodes on the top side are connected to the external readout electronics that are at virtual ground (0 V).

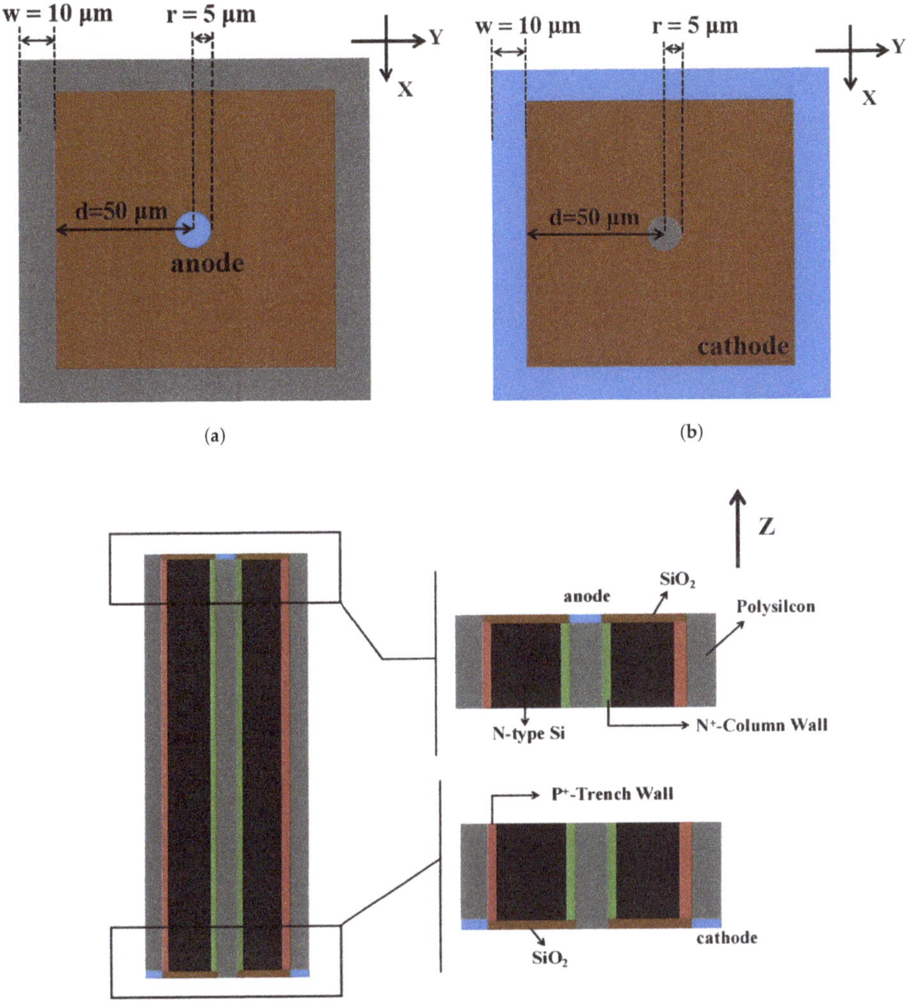

Figure 4. Structure of novel polysilicon-fill-strengthened etch-through 3D trench electrode detector structure. The axis units are micrometers. The directions of observation here are the (**a**) top side, (**b**) bottom side, and (**c**) sectional view.

3.2. Full Depletion Voltage and Electric Potential Distribution of a Novel Polysilicon-Fill-Strengthened Etch-Through 3D Trench Electrode Detector

In the novel polysilicon-fill-strengthened etch-through 3D trench electrode detector, the N-type silicon bulk has a doping concentration of $1 \times 10^{12}/\text{cm}^3$, so the effective doping concentration $N_{eff} = 1 \times 10^{12}/\text{cm}^3$. The detector thickness is chosen as 340 μm.

According to the full depletion voltage formula of traditional 3D trench electrode detectors [12]:

$$V_{fd} = \frac{eN_{eff}}{4\epsilon_{Si}\epsilon_0} d^2 + V_{bi} \tag{1}$$

where e is the electronic charge, $e = 1.6 \times 10^{19}$ C; ϵ_0 is the vacuum dielectric constant, $\epsilon_0 = 8.854 \times 10^{-12}$ F/m; ϵ_{Si} is the relative dielectric constant of silicon, $\epsilon_{Si} = 11.9$; d is the electrode spacing, d = 45 μm; and V_{bi} is the built-in potential of the p-n junction in silicon materials, $V_{bi} \approx 0.75$ V. We could obtain a full depletion voltage $V_{fd} \approx 1.5$ V, so the depletion voltage of our novel polysilicon-fill-strengthened etch-through 3D trench electrode detector should be around 1.5 V, which is close to our previous value of 1.4 V obtained from simulations.

The electron concentration distribution obtained from the simulation is presented in Figure 5. The direction of the Y-axis in Figure 5 is perpendicular to the direction of the cylindrical anode. When the bias voltage increases, the electron concentration in the detector gradually decreases until it is completely depleted. Due to the N-type heavy doping of the silicon body around the columnar electrode at the center of the detector unit, the electron concentration at this location shows a stepped upward trend along the Y-axis under different bias voltages. The simulation results show that when the bias voltage rises to 1.4 V, the carrier concentration in the active region of the detector decreases to $1 \times 10^{12}/\text{cm}^3$ below the intrinsic concentration of the silicon substrate. Therefore, we can conclude that the full depletion voltage of the detector is about 1.4 V, which has a small error compared to the result we calculated using the formula for full depletion voltage, which is about 1.5 V.

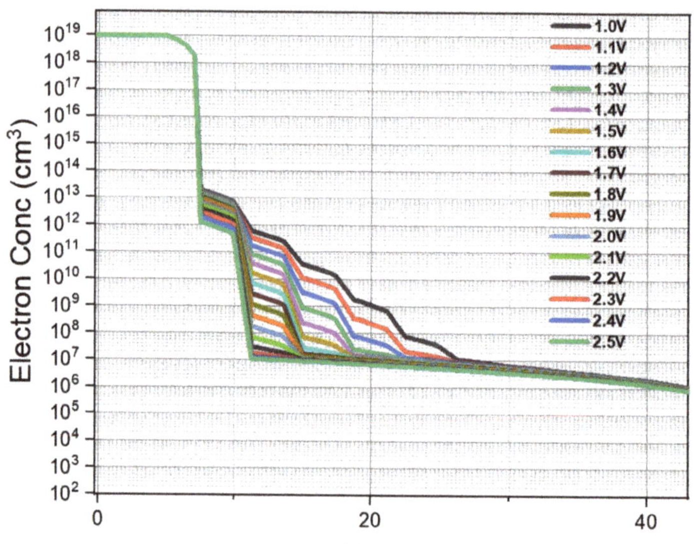

Figure 5. Partial plot of the 2D electron concentration curves of a novel polysilicon-fill-strengthened etch-through 3D trench electrode detector under different voltages.

Figure 6 illustrates the electric potential distribution for the novel polysilicon-fill-strengthened etch-through 3D trench electrode detector at different bias voltages. As shown by the Figure 6, the detector exhibits a highly symmetrical electric potential distribution. To establish reverse biasing conditions and facilitate the drift and collection of free electrons at the anode, the system employs maximum (at or over full depletion voltage) positive electrical potential at the central column electrode of the detector.

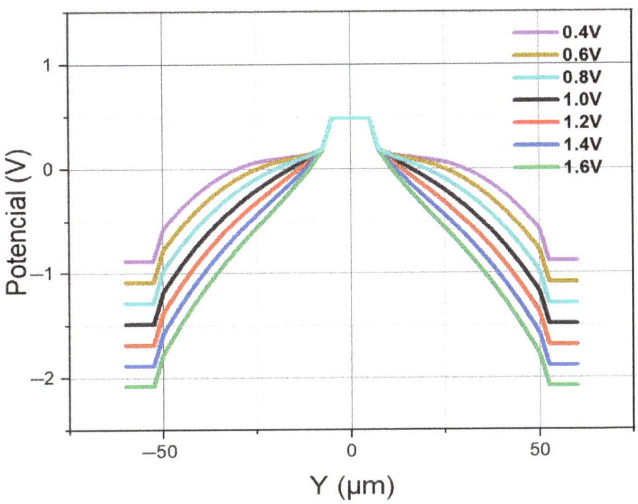

Figure 6. One-dimensional electric potential distribution of a novel polysilicon-fill-strengthened etch-through 3D trench electrode detector under different voltages.

3.3. Electric Field Distribution of a Novel Polysilicon-Fill-Strengthened Etch-Through 3D Trench Electrode Detector

The electric field distribution is an important characteristic of the novel polysilicon-fill-strengthened etch-through 3D trench electrode detector. Figure 7 shows the distribution of the electric field inside the detector along the Z-axis direction at a bias voltage of 1.4 V and Z = 170 μm. It can be seen from the figure that the electric field strength is highest in the central area of the detector, lower in the edge area, and gradually decreases from the center to the outside. The electric field distribution is symmetrical, and the electric field strength in the central area is relatively uniform.

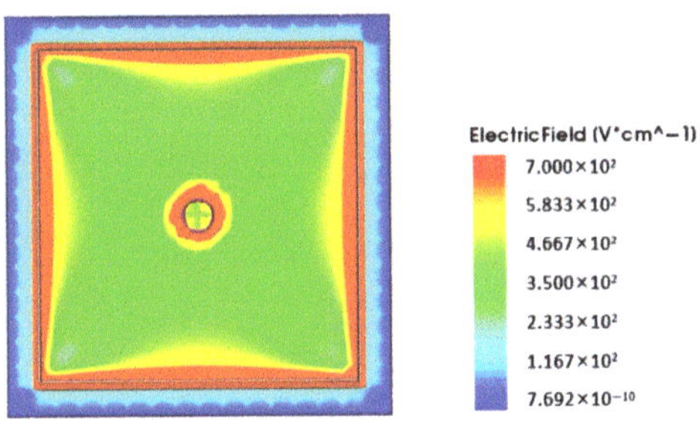

Figure 7. At a bias voltage of 1.4 V, when Z = 170 μm, the electric field distribution of a novel polysilicon-fill-strengthened etch-through 3D trench electrode detector along the Z-axis is presented.

We established a 2D cross-section at X = 0 to more clearly display the electric field distribution inside the detector. The simulation results depicting the 2D and 1D electric field distributions of the novel polysilicon-fill-strengthened etch-through 3D trench electrode detector under different bias voltages are illustrated in Figure 8a,b. According to the analysis of the graphical results, the low electric field region within the detector gradually diminishes as the bias voltage increases. When the applied voltage reaches the full depletion voltage, the internal electric field fully covers the entire detector, and the original zero-field region completely disappears. Based on the data analysis from Figure 8, it can be determined that the full depletion voltage of this detector is approximately 1.4 V, which is consistent with the analysis of the changes in electron concentration inside the detector mentioned above. In addition, it can also be clearly seen that the electric field distribution in the depleted area of the detector is extremely uniform, ensuring good charge collection throughout the entire effective detector body.

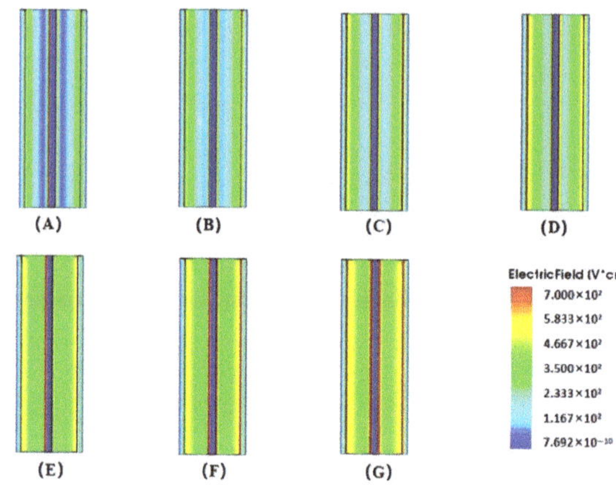

(a)

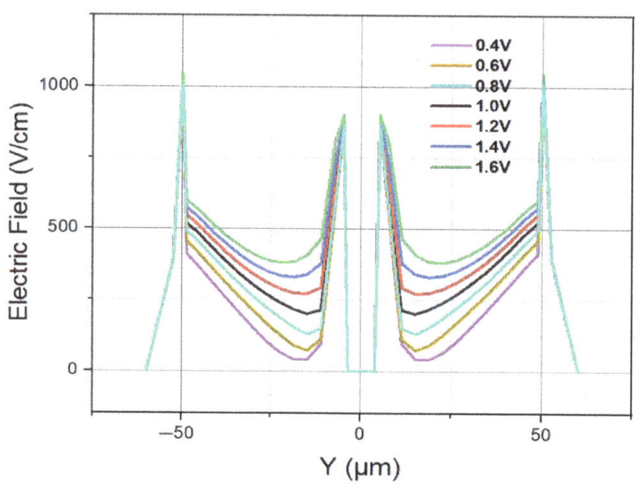

(b)

Figure 8. The (**a**) 2D and (**b**) 1D electric field distribution of a novel polysilicon-fill-strengthened etch-through 3D trench electrode detector under different voltages: (**A**) 0.4 V, (**B**) 0.6 V, (**C**) 0.8 V, (**D**) 1.0 V, (**E**) 1.2 V, (**F**) 1.4 V, (**G**) 1.6 V.

Figure 9 shows the electric field distribution of a traditional 3D trench electrode silicon detector with the same size. The electrode spacing of the traditional 3D trench detector is 50 μm and the un-etched substrate thickness is 20 μm. Comparative analysis shows that there is significant non-uniformity of the electric field in the un-etched substrate volume at the bottom of traditional 3D trench electrode silicon detectors. In contrast, the new detector achieves excellent electric field uniformity throughout the entire effective volume, which is achieved through symmetrical trench etching and optimized electrode filling. This structure ensures a uniform carrier transport path, thereby improving the charge collection efficiency of the detector in high-precision detection applications.

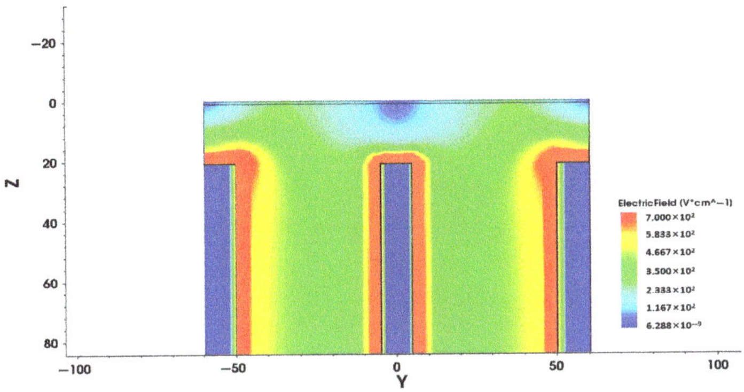

Figure 9. Two-dimensional electric field distribution of traditional 3D trench electrode silicon detector under a bias voltage of 1.4 V.

3.4. Leakage Current and Capacitance of a Novel Polysilicon-Fill-Strengthened Etch-Through 3D Trench Electrode Detector

Leakage current is the noise of the detector itself, and this noise level is closely related to the detector material structural design. By optimizing material selection and manufacturing processes (such as reducing interface defects), the leakage current of complex structures can be effectively suppressed due to the reduction in the density of defects. The magnitude of leakage current directly affects the detector performance [21].

The leakage current per unit volume caused by irradiation is directly proportional to the irradiation fluence. For a 1 MeV equivalent neutron irradiation fluence, φ_n, the change in current can be written as:

$$\frac{J}{Vol_{dep}} = \alpha\varphi_n \tag{2}$$

where α is the damage coefficient, $\alpha = 4 \times 10^{-17}\mathrm{A/cm}$, and Vol_{dep} is the volume of the depletion zone [19].

Assuming the effective defect energy level is in the middle of the energy band, the leakage current in the depletion region can be expressed as:

$$J \cong J_{gen} = \frac{en_i Vol_{dep}}{2\tau} \tag{3}$$

where e is the basic charge, n_i is the intrinsic carrier concentration in silicon ($1.5 \times 10^{10}/\mathrm{cm}^3$), and τ is the minority carrier lifetime [22].

The value of τ obtained from Equations (2) and (3) is:

$$\tau = \frac{en_i}{2\alpha\varphi_n} \tag{4}$$

and will be input into the SRH composite model in our simulation tool for calculating the leakage current of the 3D trench electrode silicon detector after irradiation in our simulations.

Figure 10 shows the simulated leakage current (with the input of Equation (4) in the simulation model of CONSRH) curves of the novel polysilicon-fill-strengthened etch-through 3D trench electrode detector before irradiation and after a radiation to a fluence of 1×10^{16} n_{eq}/cm^2, respectively. Here, we use the doping concentration to simulate the space charge induced by radiation effects. The current approximation treatment does not take into account the secondary effects that may be caused by actual deep energy levels, such as temperature dependence and the double junction effect; however, this may not affect the main conclusions of this work. As the detector is irradiated by neutrons and charged particles, the effective doping concentration N_{eff} will increase linearly with 1 MeV neutron-equivalent fluence φ_{neq} at high fluences, as (for N-type Si) [23]:

$$N_{eff} \cong N_{d0}e^{-\gamma\varphi_{eq}} - \beta\varphi_{eq} \tag{5}$$

where N_{d0} is the initial doping concentration, γ is the donor impurity removal rate, β is the deep level acceptor introduction rate, and φ_{eq} is the 1 MeV neutron-equivalent fluence. Since γ is about 1×10^{-13} cm^2 and β here is 0.01/cm for proton radiation [19], in the case of $\varphi_{eq} > 1 \times 10^{14}$ n_{eq}/cm^2,

$$N_{eff} \cong -\beta\varphi_{eq} \tag{6}$$

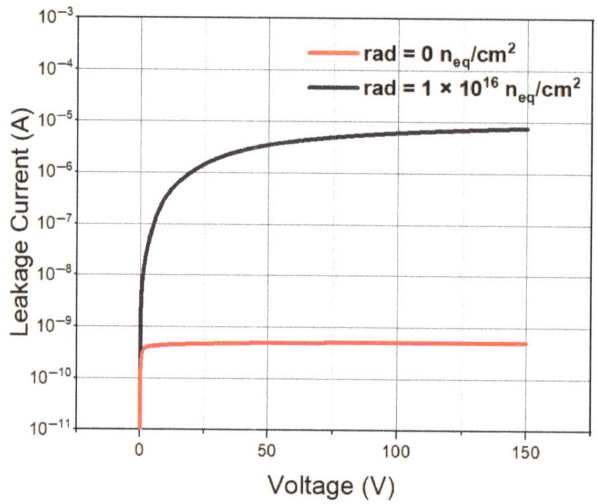

Figure 10. The leakage current curve of a novel polysilicon-fill-strengthened etch-through 3D trench electrode detector under different irradiation environments.

From Figure 10, it can be seen that as the voltage increases, the leakage current gradually increases and tends to saturate. For ease of reading, the absolute values of bias voltage and leakage current are displayed. Thus, we obtained a leakage current of 4.8×10^{-10} A for a detector unit (120 μm × 120 μm × 340 μm) with an electrode spacing of 45 μm when fully depleted. As shown in Figure 10, as the radiation fluence increases, both the detector full depletion voltage and the detector saturation leakage current increase accordingly.

In a radiation environment, the lifetime of charge carriers will be significantly shortened. As the carrier lifetime becomes shorter, the leakage current of the detector will increase, thus impacting its overall performance in a significant manner. High radiation conditions not only exacerbate this phenomenon, but also have a negative impact on the detection charge collection efficiency, response speed, and signal integrity of the detector, leading to a degradation in its performance.

As shown in Figure 11, in an environment without irradiation and low bias voltage, the capacitance of the detector is relatively large. The capacitance is inversely proportional to the depletion depth, and when the depletion depth reaches the length of that of the detector electrode spacing (in 2D cases, it is the detector thickness d, and in our 3D case it is R) the capacitance value will saturate at the theoretical value of geometric capacitance determined by the geometric structure of the detector (determined by d or R for 2D or 3D cases, respectively). Figure 11 also confirms the result obtained earlier that the full depletion voltage of the detector is around 1.4 V.

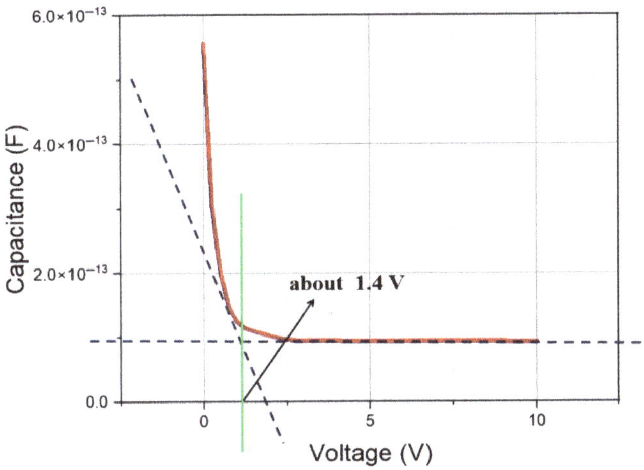

Figure 11. The capacitance curve of a novel polysilicon-fill-strengthened etch-through 3D trench electrode detector without irradiation.

According to the capacitance formula of the 3D trench electrode detector [24],

$$C = \frac{8\epsilon_r\epsilon_0}{\ln\frac{R}{r_c}} \times l \tag{7}$$

where ϵ_r is the relative dielectric constant, R is the radius of the detector, r_c is the radius of the central cylindrical electrode, and l is the length (or etching depth) of the central cylindrical electrode (and therefore it is equal to the detector thickness), the theoretical value of capacitance can be obtained as approximately 106 fF, and the simulation result of 99 fF is close to this theoretical value.

For the capacitance–voltage simulations of irradiated detectors, we used Equation (6) to determine the detector's effective doping concentration on irradiation. Figure 12 illustrates that when the bias voltage is low, the capacitance is large. As the radiation flux increases, the capacitance of the detector also increases. However, after reaching the fully depleted state, the saturation capacitance of the detector remains unchanged, and the size of the capacitor is mainly determined by the structure of the detector, namely the geometric capacitance. From the simulation results, it can be seen that as the radiation fluence increases, the full depletion voltage of the detector shows an upward trend. This indicates that the increase in radiation fluence has a significant impact on the electrical performance of the detector, leading to an increase in the full depletion voltage, and therefore the detector's operation voltage. The results demonstrate that even at a extremely high radiation fluence of 1×10^{16} n_{eq}/cm^2, the full depletion of our new detector is only about 100 Volts, which is entirely operational for the detector. We note that applying 100 V between the two electrodes may need additional insulation layers that may require extra processing steps in the fabrication of the detector.

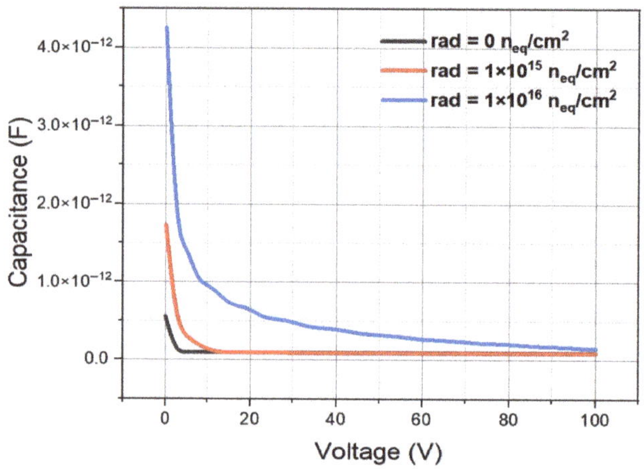

Figure 12. The capacitance curve of a novel polysilicon-fill-strengthened etch-through 3D trench electrode detector under different irradiation environments.

3.5. The Weighting Field of a Novel Polysilicon-Fill-Strengthened Etch-Through 3D Trench Electrode Detector

The detector weighting field is an important parameter for studying the coherence of silicon detector arrays, which is related to the structure of the detector and the selection of the collection electrode. Figure 13a shows the 3D distribution of the weighting field of the detector array, while the 2D distribution obtained through the cross-sectional along the X-axis is shown in Figure 13b. The findings reveal that the weighting field in the neighboring unit is nearly negligible, suggesting minimal crosstalk and very weak coherence between adjacent units. From Figure 13, it can also be seen that the electric field strength at the corners of the square unit is lower than that in the central area, which is due to the asymmetrical square shape deviated from the symmetrical circular shape. In this work, without losing generality, we choose the square shape for simplicity in the simulation process. A more symmetrical hexagonal-shaped detector structure will be studied in the future.

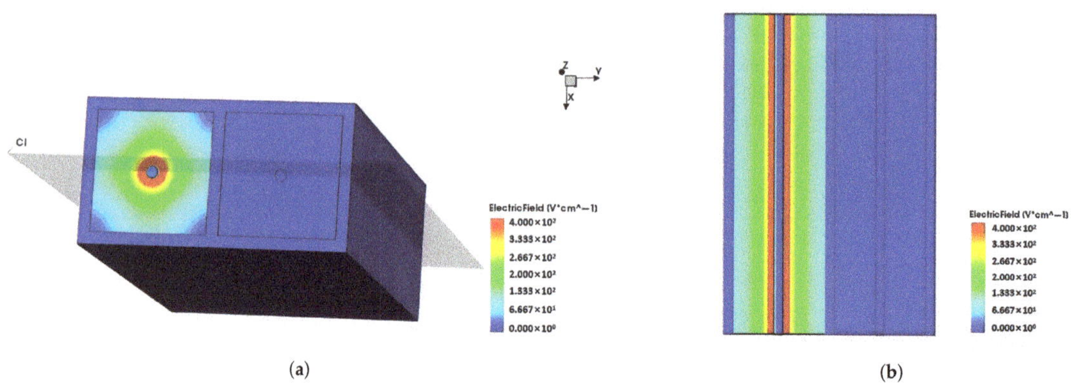

Figure 13. The (**a**) 3D and (**b**) 2D weighting field distributions of a novel polysilicon-fill-strengthened etch-through 3D trench electrode detector.

3.6. The Transient Induced Current and Charge Collection Efficiency of a Novel Polysilicon-Fill-Strengthened Etch-Through 3D Trench Electrode Detector

The simulation of transient current signal induced on the detector electrode is based on the fundamental principle of the Shockley–Ramo theorem proposed by William Shockley and Simon Ramo in 1949 [25], which is used to calculate the induced current generated when charged particles incident on to electronic devices, generating electron and hole pairs. From Figure 14, it can be seen that electric field along the diagonal directions are lower than other regions due to the deviation in symmetry of the square shape from that of the circular shape. If the incident point of a particle falls on those diagonal lines with lower electric fields, the charge collection efficiency will be lower than those of other incident points for irradiated detectors. It is worth noting that the best charge collection efficiency is obtained for particle incident points falling along the lines at $\theta = 0°, 90°, 180°$, and $270°$ [24].

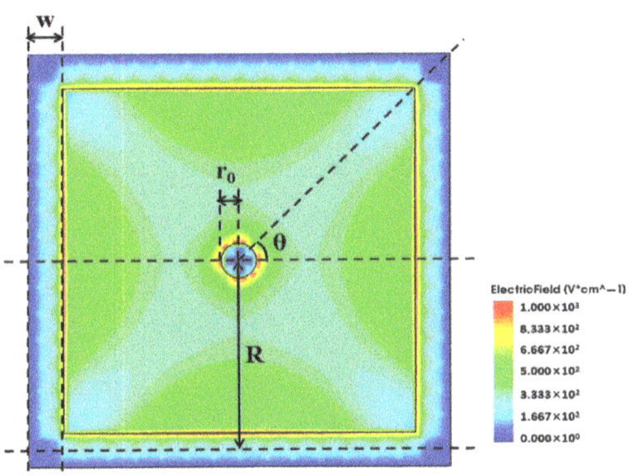

Figure 14. Two-dimensional cross-section of electric field at z = 170 under non-irradiation conditions, V = 1.4 V.

According to the Shockley–Ramo theorem, the induced current for electrons and holes at t time is [26,27]:

$$i^{e,h}(t) = q_{e,h} \cdot \overrightarrow{v_{dr}^{e,h}} \cdot \overrightarrow{E_w} \tag{8}$$

where the charge for electrons or holes q = 80 e's/μm × d_{eff} for the minimum ionization particle incident, v_{dr} is the drift velocity for electrons and holes, E_w is the weighting field, obtained from 3D simulation, and d_{eff} is the effective depth of the detector and of the trench electrode; the value of d_{eff} here is 340 μm. The carrier drift velocity v_{dr} (applicable to both electrons and holes) can be mathematically described by incorporating the saturation velocity v_s effect through the following expression:

$$v_{\text{dr}} = \frac{dr^{e,h}}{dt} = \frac{u^{e,h}E(r^{e,h})}{1 + \frac{u^{e,h}E(r^{e,h})}{v_s}} \tag{9}$$

Here, for both electrons and holes, the approximate value of v_s is about 1×10^7 cm/s, $u^{e,h}$ is the low field mobility of charge carriers, where electrons are 1450 cm^2/s/V and holes are 450 cm^2/s/V, and $E(r^{e,h})$ is the electric field strength at the position $r^{e,h}$ of the drift electron or hole, which is obtained through the three-dimensional electric field simulation of the detector.

In the simulation, we use q = 80 e's/μm (ionization charge density per unit path length) as the physical expression of calculation code of LET_S in the simulation tool. LET_S represents the rate at which particles deposit energy, i.e., linear energy transfer, in silicon in simulation language. In this study, we adopted a value of 1.28×10^{-5} for LET_S. By using the minimum ionization particle model in the TCAD, we obtained the induced current characteristics in the region of the effective silicon body of the detector with similar methods used in [28], with the results shown in Figure 15. We note that in these transient induced current simulations, Equations (8) and (9) were not actually used, which are only for the purposes of physical explaining. Figure 15 shows the relationship between the detector induced current and drift time before and after MIP incidence at the point of (X = 10, Y = 10, Z = 0) along the Z axis with a bias voltage of 100 V. From the figure, it can be seen that the peak values of the current curve are all within 100 ps, indicating that the detector's response speed is relatively fast. It can also be seen that the induced current generated inside the detector after irradiation is smaller than the induced current inside the detector under non-irradiation conditions. This is because irradiation can lead to an increase in defects, causing carriers to be trapped during the drift process, and carriers cannot all drift to the electrode, so the induced current of the detector will decrease. By integrating the induced current, we obtained the total charge (about 4 fC).

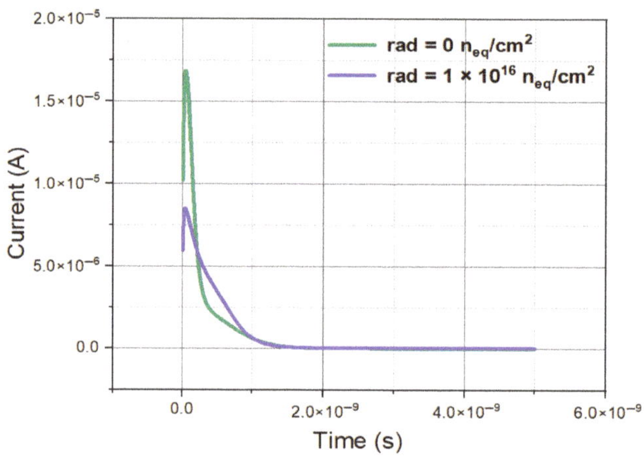

Figure 15. The induced current curve of a novel polysilicon-fill-strengthened etch-through 3D trench electrode detector under different irradiation environments.

For irradiated detectors, there will be charge trapping due to radiation-generated defects. The induced current will be reduced in time due to charge trapping [22]:

$$i(t) = q_0 \cdot v_{dr}(t) \cdot E_w \cdot e^{-\frac{t}{\tau_t}} \tag{10}$$

where q_0 is the initial charge generated by an incident particle, and τ_t is the trapping constant, ${\tau_t}^{-1} = 5 \times 10^{-7} \varphi_{eq}$. Therefore, the collected charge at time t is

$$Q(t) = \int_0^t q_0 \cdot v_{dr}(t) \cdot E_w \cdot e^{-\frac{t}{\tau_t}} dt \tag{11}$$

The charge collected for an incident induced charge moving from particle incident point r_1 to r is

$$Q(r) = q_0 \cdot \int_r^{r_1} E_w \cdot e^{-\frac{r_1 - r}{v_S \tau_t}} dr \tag{12}$$

where for large bias voltages, we have approximated $V_{dr}(t)$ as the saturation velocity $v_s = 1 \times 10^7$ cm/s, and $r = r_1 - v_s t$ ($t = \frac{r_1 - r}{v_s}$).

To estimate $Q(r)$, we further approximate the weighting field E_w as a constant, $E_w = \frac{1}{(R-r_0-\frac{w}{2})}$, without losing physical meaning (the real $E_w(r)$ is $\frac{1}{r\ln(\frac{R-\frac{w}{2}}{r_0})}$, which will significantly complicate the calculations). Equation (12) can be solved and written as:

$$Q(r) = q_0 \frac{v_s \tau_t e^{-\frac{r_1}{v_s \tau_t}}}{R - r_0 - \frac{w}{2}} \left(e^{\frac{r_1}{v_s \tau_t}} - e^{\frac{r}{v_s \tau_t}} \right) = q_0 \frac{v_s \tau_t}{R - r_0 - \frac{w}{2}} \left(1 - e^{-\frac{r_1 - r}{v_s \tau_t}} \right) \tag{13}$$

When $r = r_0$, $r_1 = R - \frac{w}{2}$, the collected charge is:

$$Q = q_0 \frac{v_s \tau_t}{R - r_0 - \frac{w}{2}} \left(1 - e^{-\frac{R - r_0 - \frac{w}{2}}{v_s \tau_t}} \right) \tag{14}$$

Therefore, the charge collection efficiency is

$$\text{CCE} = \frac{Q}{q_0} = \frac{v_s \tau_t}{R - r_0 - \frac{w}{2}} \left(1 - e^{-\frac{R - r_0 - \frac{w}{2}}{v_s \tau_t}} \right) \tag{15}$$

For our case, R = 55 μm, r_0 = 5 μm, w = 10 μm, at $\varphi_{eq} = 1 \times 10^{16}$ $\mathrm{n_{eq}/cm^2}$, $v_s \tau_t$ = 20 μm, CCE = 39.76%. Therefore, our detector is quite radiation-hard. In reference [24], it was noted by the researchers that charge collection efficiency remains improvable through further minimization of the electrode spacing.

We noticed that there was a recent report on a 3D trench wall electrode (straight) detector showing that even with a extremely high radiation fluence of 1×10^{17} $\mathrm{n_{eq}/cm^2}$, there was a good timing resolution of 10 ps [29].

3.7. Detector Array of a Novel Polysilicon-Fill-Strengthened Etch-Through 3D Trench Electrode Detector

The novel polysilicon-fill-strengthened etch-through 3D trench electrode detector unit can form a detector array with ultra-fast and ultra-radiation-resistant characteristics. Figure 16a,b is a schematic diagram of a 3 × 3 array, where adjacent detector units are separated by highly doped trenches and share the same trench electrodes, greatly reducing dead zones and significantly improving detector performance. Each detector pixel is connected to a readout channel, which is connected to an application-specific integrated circuit (ASIC).

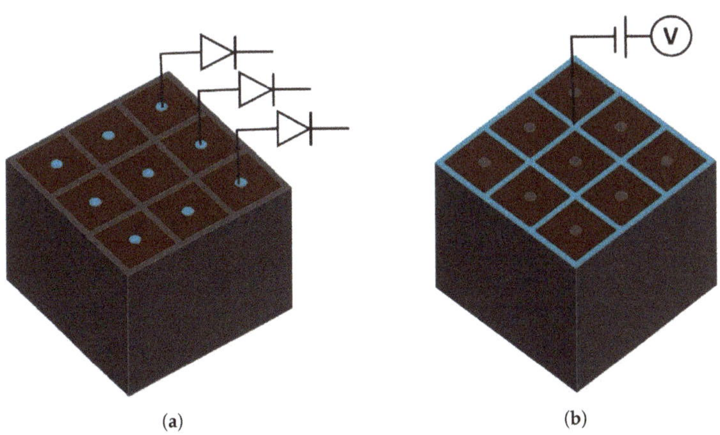

Figure 16. The schematic of (**a**) the top side and (**b**) the bottom side of a novel polysilicon-fill-strengthened etch-through 3D trench electrode detector array.

4. Conclusions

After in-depth research on the design and functionality of a new type of 3D trench electrode detector, we conducted detailed and comprehensive simulation studies using Sentaurus TCAD simulation software. We obtained key electrical parameters such as the electron concentration profiles, full depletion voltage, electric potential distribution, electric field distribution, leakage current, capacitance, weighting field distribution, and the induced transient current of the detector. In addition, we also conducted simulation tests and analyses on detectors that had undergone high-fluence irradiation to evaluate their long-term stability and performance degradation.

This new type of detector is etched and filled on both sides of the silicon detector to form through-trench electrodes, which solves the problem of the detector's main body falling off the detector. Separating the biasing electrode and readout electrode on two opposite sides of the detector may prevent breakdown between the two electrodes caused by discharge between the metal contacts of the two electrodes. In the absence of irradiation, the full depletion voltage of our detector is approximately 1.4 V, which is close to the theoretical calculated value. Our detector has a low capacitance, which is a common characteristic of 3D trench electrode detectors. Calculation results show that our detector can sustain a radiation fluence of up to 1×10^{16} n_{eq}/cm^2, with a CCE of about 40%.

For pixillated detectors, the simulation results show that our units in the detector are isolated from each other due to etch-through highly doped trenches, eliminating the electronic interference between them.

Author Contributions: Conceptualization, X.Z. and Z.L. (Zheng Li); data curation, X.Z.; formal analysis, Z.L. (Zheng Li); funding acquisition, Z.L. (Zheng Li); investigation, X.Z., T.L., J.Z., and X.L.; methodology, X.Z.; project administration, Z.L. (Zheng Li); resources, X.Z., Z.L. (Zheng Li), and M.W.; software, X.Z., Z.L. (Zhiyu Liu), T.L., and J.Z.; supervision, X.Z., Z.L. (Zheng Li), J.Z., X.L., M.L., and M.W.; validation, X.Z., Z.L. (Zheng Li), Z.L. (Zhiyu Liu), T.L., J.Z., X.L., and M.W.; visualization, X.Z. and Z.L. (Zheng Li); writing—original draft, X.Z.; writing—review and editing, X.Z., Z.L. (Zheng Li), and Z.L. (Zhiyu Liu). All authors have read and agreed to the published version of the manuscript.

Funding: This research was funded by the National Key R&D Program of China (2023YFF0719600) and the Key Project of the National Natural Science Foundation of China (11835008).

Data Availability Statement: The original contributions presented in the study are included in the article, further inquiries can be directed to the corresponding author.

Conflicts of Interest: The authors declare no conflicts of interest.

References

1. Stapnes, S. Detector challenges at the LHC. *Nature* **2007**, *448*, 290–296. [CrossRef] [PubMed]
2. Ruzin, A. Review of semiconductor ionizing radiation detectors. *Nat. Sci. J. Xiangtan Univ.* **2018**, *4*, 106–114.
3. Galloway, Z.; Gee, C.; Mazza, S.; Ohldag, H.; Rodriguez, R.; Sadrozinski, H.F.-W.; Schumm, B.A.; Seiden, A.; Wyatt, W.; Zhao, Y. Use of 'LGAD' ultra-fast silicon detectors for time-resolved low-keV X-ray science. *Nucl. Instrum. Methods Phys. Res.* **2019**, *923*, 5–7. [CrossRef]
4. Meier, D.; Czermak, A.; Jalocha, P.; Sowicki, B.; Kowal, M.; Dulinski, W.; Maehlum, G.; Nygard, E.; Yoshioka, K.; Fuster, J.; et al. Silicon detector for a Compton camera in nuclear medical imaging. *IEEE Trans. Nucl. Sci.* **2002**, *49*, 812–816. [CrossRef]
5. Pourmorteza, A.; Symons, R.; Sandfort, V.; Mallek, M.; Fuld, M.K.; Henderson, G.; Jones, E.C.; Malayeri, A.A.; Folio, L.R.; Bluemke, D.A. Abdominal Imaging with Contrast-enhanced Photon-counting CT: First Human Experience *Radiology* **2016**, *279*, 239–245. [CrossRef]
6. Granfors, P.R.; Aufrichtig, R.; Possin, G.E.; Giambattista, B.W.; Huang, Z.S.; Liu, J.; Ma, B. Performance of a 41 × 41 cm^2 amorphous silicon flat panel x-ray detector for radiographic imaging applications. *Med. Phys.* **2000**, *27*, 1324–1331. [CrossRef]
7. Shi, L.; Sarubbi, F.; Nihtianov, S.N.; Nanver, L.K.; Scholtes, T.L.M.; Scholze, F. High performance silicon-based extreme ultraviolet (EUV) radiation detector for industrial application. In Proceedings of the 2009 IECON'09 35th Annual Conference of IEEE Industrial Electronics, Porto, Portugal, 3–5 November 2009; IEEE: Piscataway, NJ, USA, 2009; pp. 1877–1882.

8. Kemmer, J. Fabrication of low noise silicon radiation detectors by the planar process. *Nucl. Instrum. Methods* **1980**, *169*, 499–502. [CrossRef]
9. Maisch, T.; Günzler, R.; Weiser, M.; Kalbitzer, S.; Welser, W.; Kemmer, J. Ion-implanted Si pn-junction detectors with ultrathin windows. *Nucl. Instrum. Methods Phys. Res. Sect. A Accel. Spectrometers Detect. Assoc. Equip.* **1990**, *288*, 19–23. [CrossRef]
10. Gatti, E.; Rehak, P. Semiconductor drift chamber—An application of a novel charge transport scheme. *Nucl. Instrum. Methods Phys. Res.* **1984**, *225*, 608–614. [CrossRef]
11. Guazzoni, C. The first 25 years of silicon drift detectors: A personal view. *Nucl. Instrum. Methods Phys. Res. Sect. A Accel. Spectrometers Detect. Assoc. Equip.* **2010**, *624*, 247–254. [CrossRef]
12. Parker, S.; Kenney, C.; Segal, J. 3D—A proposed new architecture for solid-state radiation detectors. *Nucl. Instrum. Methods Phys. Res. Sect. A Accel. Spectrometers Detect. Assoc. Equip.* **1997**, *395*, 328–343. [CrossRef]
13. Kenney, C.; Parker, S.; Segal, J.; Storment, C. Silicon detectors with 3D electrode arrays: Fabrication and initial test results. *IEEE Trans. Nucl. Sci.* **1999**, *46*, 1224–1236. [CrossRef]
14. DaVia, C.; Hasia, J.; Kenney, C.; Koka, A.; Parker, S. 3D silicon detectors—Status and applications. *Nucl. Instrum. Methods Phys. Res. Sect. A Accel. Spectrometers Detect. Assoc. Equip.* **2005**, *549*, 122–125. [CrossRef]
15. Li, Z. New BNL 3D-Trench electrode Si detectors for radiation hard detectors for sLHC and for X-ray applications. *Nucl. Instrum. Methods Phys. Res. A* **2011**, *658*, 90–97. [CrossRef]
16. Montalbano, A.; Bassignana, D.; Li, Z.; Liu, S.; Lynn, D.; Pellegrini, G.; Tsybychev, D. A systematic study of BNL's 3D-Trench Electrode detectors. *Nucl. Instrum. Methods Phys. Res. A* **2014**, *765*, 23–28. [CrossRef]
17. Kenney, C.; Parker, S.; Walckiers, E. Results from 3D silicon sensors with wall electrodes: Near-cell-edge sensitivity measurements as a preview of active-edge sensors. *IEEE Trans. Nucl. Sci.* **2001**, *48*, 2405–2410. [CrossRef]
18. Betta, G.F.D.; Bagolini, A.; Boscardin, M.; Bosisio, L. Development of modified 3D detectors at FBK. In Proceedings of the IEEE Nuclear Science Symposium & Medical Imaging Conference, Knoxville, TN, USA, 30 October–6 November 2010; ISBN 9781424491049.
19. Liu, M.; Jiang, S.; Cheng, W.; Ji, H. A double-sided 3D trench electrode detector using an 8-inch CMOS process: 3D simulation and experimental investigation. *Nucl. Instrum. Methods Phys. Res. A* **2025**, *10170*, 169943. [CrossRef]
20. Kraner, H.; Li, Z.; Posnecker, K. Fast neutron damage in silicon detectors. *Nucl. Instrum. Methods Phys. Res. A Accel. Spectrometers Detect. Assoc. Equip.* **1989**, *279*, 266–271. [CrossRef]
21. Wu, C.; Huang, X.; Xiao, Y.; Li, Z. Design and simulation of a new silicon pixel detector with equal width floating electrodes. *AIP Adv.* **2021**, *11*, 025315. [CrossRef]
22. Schroder, D.K. *Semiconductor Material and Device Characterization*; John Wiley & Sons: Hoboken, NJ, USA, 2006.
23. Dezillie, B.; Li, Z.; Eremin, V.; Chen, W.; Zhao, L. The effect of oxygen impurities on radiation hardness of FZ silicon detectors for HEP after neutron, proton and gamma irradiation. *IEEE Trans. Nucl. Sci.* **2000**, *47*, 1892–1897. [CrossRef]
24. Ding, H.; Chen, J.; Li, Z.;Yan, S. Modeling and simulation of charge collection properties for 3D-trench electrode detector. *Nucl. Instrum. Methods Phys. Res. A* **2015**, *796*, 29–33. [CrossRef]
25. Liu, M.; Zhou, T.; Li, Z. Electrical properties of ultra-fast 3D-Trench Electrode silicon detector. *Micromachines* **2020**, *11*, 674. [CrossRef]
26. Lemeilleur, F.; Glaser, M.; Heijne, E.; Jarron, P.; Occelli, E.; Rioux, J. Electrical properties and charge collection efficiency for neutron-irradiated p-type and n-type silicon detectors. *Nucl. Phys. B-Proc. Suppl.* **1993**, *32*, 415–424. [CrossRef]
27. Li, Z.; Kraner, H. Modeling and simulation of charge collection properties for neutron irradiated silicon detectors. *Nucl. Phys. B-Proc. Suppl.* **1993**, *32*, 398–409. [CrossRef]
28. Tan, Z.; Li, Z.; Li, X.; Cai, X.; Liu, M.; Wang, H. Study of electrical properties and detection mechanism of a practical novel 3D-Spherical Electrode Detector. *Front. Mater.* **2023**, *10*, 1149774. [CrossRef]
29. Lampis, A.; Addison, M.; Bellora, A.; Boscardin, M.; Brundu, D.; Cardini, A.; Cossu, G.M.; Betta, G.F.D.; La Delfa, L.; Lai, A.; et al. Performance of 3D trench silicon pixel sensors irradiated up to 1×10^{17} 1 MeV n_{eq} cm^{-2}. *Nucl. Instrum. Methods Phys. Res. A* **2024**, *1069*, 169984. [CrossRef]

micromachines

MDPI

Article

Highly Efficient Inverted Organic Light-Emitting Devices with Li-Doped MgZnO Nanoparticle Electron Injection Layer

Hwan-Jin Yoo, Go-Eun Kim, Chan-Jun Park, Su-Been Lee, Seo-Young Kim and Dae-Gyu Moon *

Department of Electronic Materials, Device, and Equipment Engineering, Soonchunhyang University, Asan-si 31538, Republic of Korea; yhjin1209@sch.ac.kr (H.-J.Y.); tkfkdrhdms153@sch.ac.kr (G.-E.K.); cjpark5581@sch.ac.kr (C.-J.P.); 20247114@sch.ac.kr (S.-B.L.); syoung031219@sch.ac.kr (S.-Y.K.)

* Correspondence: dgmoon@sch.ac.kr; Tel.: +82-41-530-1312

Abstract: Inverted organic light-emitting devices (OLEDs) have been attracting considerable attention due to their advantages such as high stability, low image sticking, and low operating stress in display applications. To address the charge imbalance that has been known as a critical issue of the inverted OLEDs, Li-doped MgZnO nanoparticles were synthesized as an electron-injection layer of the inverted OLEDs. Hexagonal wurtzite-structured Li-doped MgZnO nanoparticles were synthesized at room temperature via a solution precipitation method using LiCl, magnesium acetate tetrahydrate, zinc acetate dihydrate, and tetramethylammonium hydroxide pentahydrate. The Mg concentration was fixed at 10%, while the Li concentration was varied up to 15%. The average particle size decreased with Li doping, exhibiting the particle sizes of 3.6, 3.0, and 2.7 nm for the MgZnO, 10% and 15% Li-doped MgZnO nanoparticles, respectively. The band gap, conduction band minimum and valence band maximum energy levels, and the visible emission spectrum of the Li-doped MgZnO nanoparticles were investigated. The surface roughness and electrical conduction properties of the Li-doped MgZnO nanoparticle films were also analyzed. The inverted phosphorescent OLEDs with Li-doped MgZnO nanoparticles exhibited higher external quantum efficiency (EQE) due to better charge balance resulting from suppressed electron conduction, compared to the undoped MgZnO nanoparticle devices. The maximum EQE of 21.7% was achieved in the 15% Li-doped MgZnO nanoparticle devices.

Keywords: OLED; inverted structure; Li-doped MgZnO nanoparticles; electron injection layer

1. Introduction

Organic light-emitting devices (OLEDs) have revolutionized the display and lighting industries with their excellent characteristics such as high contrast ratio, wide viewing angles, fast response time, and thin thickness [1–4]. Additionally, OLEDs enable the implementation of transparent, flexible, stretchable, and lightweight displays, making them a highly promising next generation display technology [5,6]. Among OLED technologies, active-matrix OLEDs (AMOLEDs) have been widely used as a key display technology for smartphones, televisions, and monitors due to their low power consumption and high image quality [7,8]. AMOLEDs utilize thin-film transistors (TFTs) to control each pixel, enabling high-speed switching and superior power efficiency compared to passive-matrix OLEDs. Particularly, oxide TFTs are gaining significant attention because they provide high mobility, a simplified manufacturing process, and high device stability [9–11]. Since oxide TFTs primarily use n-type oxide semiconductors, there is growing interest in inverted

 https://doi.org/10.3390/mi16060617

OLEDs, which can mitigate luminance degradation, operational stress, and image sticking issues [12–14].

Conventional OLEDs have a structure in which the hole-injecting electrode is located at the bottom and the electron-injecting electrode is positioned at the top of the devices. In contrast, inverted OLEDs adopt an opposite structure, where the cathode is placed at the bottom and the anode at the top. This structure offers advantages such as improved environmental stability, reduced sensitivity to air and moisture, solution processing capability as well as better compatibility with n-type oxide TFT backplanes [15–19]. A major issue in inverted OLEDs is electron injection from the cathode into the organic layer. In inverted OLEDs, indium tin oxide (ITO) is commonly used as a transparent cathode due to its excellent optical transmittance and electrical conductivity [19]. However, ITO has a high work function, making electron injection from the ITO cathode into the organic layer difficult [13–19]. Therefore, it is important to use a material with efficient electron injection properties as the electron injection layer.

ZnO nanoparticles have been studied to utilize the electron injection layer in the OLEDs because they have high electron mobility, high transparency in visible range, and a low conduction band minimum energy [20–22]. Since ZnO nanoparticles are generally dispersed in organic solvents such as methanol and ethanol, solution processes such as spin coating and inkjet printing are typically used to prepare ZnO nanoparticle layer. There have been very few reports on conventional OLEDs using ZnO nanoparticles because the organic solvents in the ZnO solution can damage the underlying organic layers leading to device failure [23–25]. In contrast, in an inverted structure, a ZnO nanoparticle layer is formed before coating the organic layers, preventing the organic layers from being damaged by ZnO solution. Therefore, the inverted structure has been predominantly used for ZnO nanoparticles [26–30]. In inverted structure, ZnO nanoparticles are coated onto the ITO cathode, allowing improvements in electron mobility, electrical conductivity, and surface morphology through the processes such as thermal annealing and surface treatment [27,28]. In the fabrication of inverted OLEDs, an additional interlayer such as polyethyleneimine, thin Ag nanoparticle, and $Ba(OH)_2$ layer was inserted between the ZnO nanoparticle layer and the electron transport layer to adjust electron injection [31–35]. However, this interlayer increases the complexity of the device fabrication process.

In this paper, we report on inverted OLEDs without an additional interlayer using Li doped MgZnO nanoparticles. Alkali metal-doped ZnO and MgZnO nanoparticles have been widely used in quantum dot light-emitting diodes to control electron injection and transport properties [36–39]. However, there are very few reports on inverted OLEDs utilizing alkali metal-doped ZnO and MgZnO nanoparticles [40,41]. Manzhi et al. fabricated inverted OLEDs using LiZnO and MgZnO nanoparticles [40,41]. They incorporated LiZnO nanoparticles or MgZnO nanostructures into poly[9,9-dioctylfluorenyl-2,7-diyl] (PFO) to form the nanocomposite emitting layer for the inverted OLEDs. They investigated the current-voltage characteristics of the inverted devices with PFO-LiZnO or PFO-MgZnO nanocomposites. However, they did not investigate the luminance or efficiency characteristics of the inverted devices. In addition, there is no report on the phosphorescent inverted devices with a Li-doped MgZnO nanoparticle layer. In this paper, we used the Li-doped MgZnO nanoparticle layer as an electron injection layer in the inverted phosphorescent devices. Li-doped MgZnO nanoparticles influence electron injection and charge balance in the inverted devices, resulting in highly efficient inverted devices with a maximum external quantum efficiency of 21.7%.

2. Materials and Methods

Li-doped MgZnO nanoparticles were synthesized using the solution precipitation method [37]. Magnesium acetate tetrahydrate (≥99%, Sigma Aldrich, St. Louis, MO, USA), lithium chloride (LiCl, ≥99%, Sigma Aldrich), zinc acetate dihydrate (≥99%, Sigma Aldrich), dimethyl sulfoxide (DMSO, ≥99%, Sigma Aldrich), tetramethylammonium hydroxide pentahydrate (TMAH, ≥97%, Sigma Aldrich), and ethyl alcohol were used to synthesize the Li-doped MgZnO nanoparticles. Magnesium acetate tetrahydrate was added to 30 mL of DMSO and stirred at 600 rpm until completely dissolved. LiCl was added to the magnesium acetate tetrahydrate solution and stirred at 600 rpm until completely dissolved. Then, zinc acetate dihydrate was added to the mixed solution of magnesium acetate tetrahydrate and LiCl, and stirred until completely dissolved. An amount of 4.62 mmol of TMAH was completely dissolved in 10 mL of ethanol using a vertex meter. The TMAH solution was added dropwise to the mixed solution of magnesium acetate tetrahydrate, LiCl, and zinc acetate dihydrate with continuous stirring. Then, ethyl acetate was used to precipitate Li-doped MgZnO nanoparticles. After collecting the precipitate using a centrifuge, it was washed repeatedly with ethanol to minimize residual ethyl acetate. Finally, the collected Li-doped MgZnO nanoparticles were redispersed in ethanol. The concentration of Mg in Li-doped MgZnO nanoparticles was fixed at 10%, while the concentration of Li was varied to 10% and 15% by adjusting the amounts of LiCl and zinc acetate dihydrate. For example, 0.15 mmol of LiCl and 1.275 mmol of zinc acetate dihydrate were used to synthesize 10% Li-doped MgZnO nanoparticles. The crystal structure of Li-doped MgZnO nanoparticles was investigated using X-ray diffraction method (Miniflex 600, Rigaku, Tokyo, Japan). Transmission electron microscopy (TEM, JEM-ARM200F, JEOL, Tokyo, Japan) was used to study the particle size and shape of Li-doped MgZnO nanoparticles. The UV-vis absorption spectrum of the nanoparticles was measured using a UV-vis spectrometer (Lamda 35, PerkinElmer, Waltham, MA, USA). The valence band maximum energy of Li-doped MgZnO nanoparticles was measured using ultraviolet photoelectron spectroscopy (UPS, NEXSA, Thermo Fisher Scientific, Waltham, MA, USA). Atomic force microscopy (AFM, XE7, Park Systems, Suwon, Republic of Korea) was utilized to measure the surface roughness of thin films coated with Li-doped MgZnO nanoparticles.

Inverted phosphorescent OLEDs using the synthesized Li-doped MgZnO nanoparticles were fabricated on the ITO-coated glass substrates. The sheet resistance of the ITO film was about 10 Ω/sq. ITO cathode patterns were defined using a standard photolithography process. The ITO cathode patterns were cleaned with acetone, isopropyl alcohol, and deionized water. The cleaned ITO patterns were exposed to oxygen plasma to remove organic residues before spin-coating with Li-doped MgZnO nanoparticles in ethanol. The thickness of the Li-doped MgZnO nanoparticle layer was fixed to be 10 nm. After forming the Li-doped MgZnO nanoparticle layer, organic and metal layers were deposited by the vacuum thermal evaporation method at a base pressure of 10^{-6} Torr. A 25 nm thick 2,2′,2″-(1,3,5-benzinetriyl)-tris-(1-phenyl)-1-H-benzimidazole (TPBi) electron transport layer was deposited, followed by a 20 nm thick 4,4′-bis(9-carbazolyl)biphenyl (CBP) host layer co-deposited with a green phosphorescent dopant, tris(2-phenylpyridine)iridium(III) [$Ir(ppy)_3$], for the emission layer. The concentration of $Ir(ppy)_3$ was 5 wt%. After depositing the CBP:$Ir(ppy)_3$ emission layer, a 30 nm-thick undoped CBP hole transport layer was deposited, followed by a 10 nm-thick MoO_3 hole injection layer was deposited. Finally, a 100 nm thick Al layer was evaporated to define the anode through a shadow mask without breaking the vacuum. The completed device structure is ITO/Li-doped MgZnO (10 nm)/TPBi (25 nm)/CBP:$Ir(ppy)_3$ (20 nm, 5 wt%)/CBP (30 nm)/MoO_3 (10 nm)/Al. Current density-voltage-luminance (J-V-L) characteristics of the inverted devices were measured using computer-controlled Keithley 2400 source-measure units and a calibrated

Si photodiode. The electroluminescence spectra of the device were measured using a spectroradiometer (CS1000, Minolta, Tokyo, Japan).

3. Results and Discussion

Li-doped MgZnO nanoparticles were prepared by solution precipitation method using LiCl, magnesium acetate tetrahydrate, and TMAH for the electron injection layer in inverted OLEDs. The concentration of Mg was fixed at 10%, and the concentration of Li was varied to be 10% and 15% in the Li-doped MgZnO nanoparticles. Figure 1 shows the X-ray diffraction patterns for the synthesized Li-doped MgZnO nanoparticles. The undoped MgZnO nanoparticles exhibited diffraction peaks at 2θ values of 31.6°, 34.2°, 36.0°, 47.3°, 56.4°, 62.5°, and 67.8°, which correspond to the (100), (002), (101), (102), (110), (103), and (112) planes of the hexagonal wurtzite structure of ZnO (JCPDS card 361451). The X-ray diffraction pattern of the MgZnO indicates that doping with 10% Mg into ZnO nanoparticles does not alter the hexagonal wurtzite structure of ZnO [42–44]. Even with 10% and 15% Li doping, diffraction peaks corresponding to the (100), (101), (102), (110), (103), and (112) planes of the hexagonal wurtzite structure of ZnO are observed at diffraction angles (2θ) of 31.6°, 36.1°, 47.3°, 56.4°, 62.4°, 67.9°, respectively. As shown in the figure, the X-ray diffraction pattern remains unchanged with Li doping into MgZnO, while the diffraction peaks become broader. This result indicates that the hexagonal wurtzite structure of MgZnO is maintained with Li doping, whereas the particle size of Li-doped MgZnO nanoparticles is smaller than that of the undoped MgZnO nanoparticles [45].

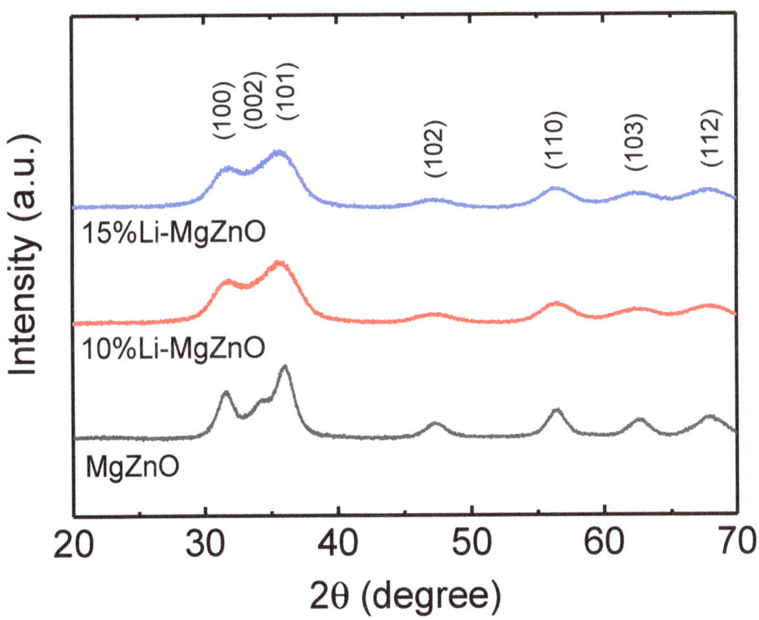

Figure 1. X-ray diffraction patterns of the Li-doped MgZnO nanoparticles.

Figure 2 shows the TEM images and the particle size distributions of the synthesized Li-doped MgZnO nanoparticles. MgZnO nanoparticles have an elliptical shape, and the Li-doped MgZnO nanoparticles also retain this elliptical morphology, indicating that Li-doping into the MgZnO nanoparticles does not cause significant changes in particle shape. The average particle size of MgZnO nanoparticles was 3.6 nm, while it decreased to 3.0 nm for 10% Li-doped MgZnO and to 2.7 nm for 15% Li-doped MgZnO nanoparticles, indicating that the particle size decreases with increasing Li doping. The particle size distribution with Li doping also shows a similar trend to that of the particle size. MgZnO nanoparticles show the highest count in the particle size range of 3.5 to 4.0 nm. With Li doping, the dominant count shifts toward smaller particle sizes. The 10% Li-doped MgZnO nanoparticles show the highest count in the particle size range of 2.8 to 3.3 nm, while the 15% Li-doped MgZnO nanoparticles exhibit the highest count in the range of 2.3 to 2.8 nm. The Li-doped MgZnO nanoparticles also exhibit a more uniform particle size compared to the MgZnO nanoparticles. MgZnO nanoparticles have particle sizes ranging from 2.0 to 5.5 nm, whereas the 10% Li-doped MgZnO nanoparticles exhibit particle sizes between 1.8 and 4.8 nm, and the 15% Li-doped MgZnO nanoparticles have particle sizes in the range of 1.8 to 4.3 nm. For the synthesis of MgZnO nanoparticles, magnesium acetate tetrahydrate, zinc acetate dihydrate, and TMAH were used as the Mg and Zn precursors and the oxidizing agent, respectively. Mg^{2+} and Zn^{2+} ions from magnesium acetate tetrahydrate and zinc acetate dihydrate react with OH^- ions from TMAH to form MgZnO nanoparticles. The MgZnO nanoparticles grow through the Ostwald ripening process [46,47]. For the synthesis of Li-doped MgZnO nanoparticles, LiCl was used as the Li precursor. It is assumed that Li^+ ions from LiCl are adsorbed onto the surface of the MgZnO particles, thereby interfering with the recombination of Mg^{2+} and Zn^{2+}, and OH^- ions. As a result, the growth rate of the nanoparticles decreases, leading to a reduction in particle size. From Figures 1 and 2, it can be observed that Li doping in MgZnO nanoparticles does not alter the crystal structure, whereas the particle size decreases. MgZnO is thermodynamically stable with a hexagonal wurtzite structure. The ionic radius of Zn^{2+} is approximately 0.74 Å, while that of Li^+ is 0.59 Å [48]. Li^+ ions can either substitute for Zn^{2+} ions or occupy interstitial sites, causing only minimal lattice distortion within the solubility limit. Therefore, the crystal structure remains unchanged by Li doping [41,49]. On the other hand, Li^+ ions readily adsorb onto the surface of MgZnO nanoparticles during crystal growth. These adsorbed Li^+ ions hinder the approach of Zn^{2+} ions to the nanoparticle surface and suppress the reaction between Zn^{2+} and OH^- ions, thereby reducing the crystal growth rate [46,47]. As a result, the particle size decreases.

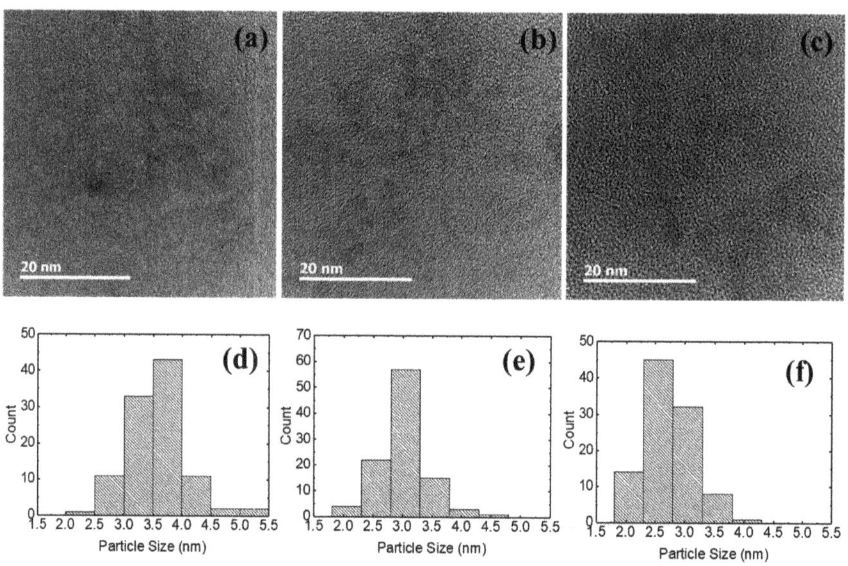

Figure 2. TEM images and particle size distributions of (**a**,**d**) MgZnO nanoparticles, (**b**,**e**) 10% Li-doped MgZnO nanoparticles, and (**c**,**f**) 15% Li-doped MgZnO nanoparticles.

Figure 3 shows the UV absorption spectra of the Li-doped MgZnO nanoparticles. As the Li concentration increases, the absorption spectrum shifts toward shorter wavelengths. While the absorption edge of MgZnO is at 347 nm, those of the 10% and 15% Li-doped MgZnO nanoparticles are at 339 and 337 nm, respectively. The blue shift of the absorption spectrum indicates an increase in the band gap. The band gap energy was determined from the intersection point between the tangent to the linear section of the absorption spectrum and the wavelength axis [20]. The measured band gaps of MgZnO, 10% and 15% Li-doped MgZnO nanoparticles are 3.57, 3.66, and 3.68 eV, respectively. This increase in the band gap is believed to be related to the reduction in particle size caused by Li-doping as shown in Figure 2. Li-doped MgZnO nanoparticles with sizes ranging from 2.7 to 3.6 nm exhibit quantum confinement effects, leading to an increase in band gap as the particle size decreases [50–53]. In addition, an increase in the band gap has been observed when Li is doped into bulk ZnO [54,55]. Similarly, it can be suggested that Li doping into MgZnO nanoparticles appears not only to increase the band gap due to a reduction in particle size but also contribute to an additional increase in band gap caused by the Li doping itself.

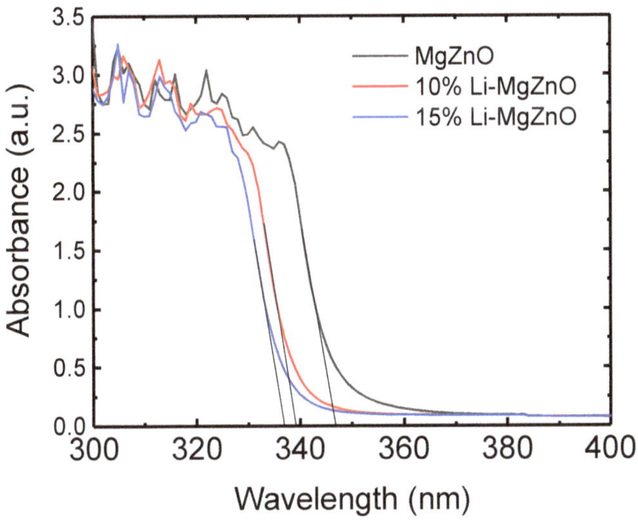

Figure 3. UV-vis absorption spectra of the Li-doped MgZnO nanoparticles.

Figure 4 shows the UPS spectra corresponding to the secondary cut-off and the valence band edge region for MgZnO, 10% Li-doped MgZnO, and 15% Li-doped MgZnO nanoparticles. The valence band maximum (VBM) was determined using the equation VBM = 21.22 − (E_{cutoff} − E_{onset}), where 21.22 eV is the incident photon energy, E_{cutoff} is the high binding energy cutoff, and E_{onset} is the onset energy of the valence band region [56]. The VBM energies for MgZnO, 10% Li-doped MgZnO, and 15% Li-doped MgZnO are 6.49, 6.59, and 6.69 eV, respectively. The VBM energy increases with Li doping. The position of the conduction band minimum (CBM) was calculated using the equation, $E_c = E_v - E_g$, where E_c is the CBM energy, E_v is the VBM energy obtained from UPS spectra, and E_g is the band gap determined from the UV-vis absorption spectrum. The CBM energies for MgZnO, 10% Li-doped, and 15% Li-doped MgZnO are 2.92, 2.93, and 3.01 eV, respectively, showing a slight increase with Li doping. The increase in CBM energy indicates a higher energy barrier for electron injection from the Li-doped MgZnO to the organic electron transport layer.

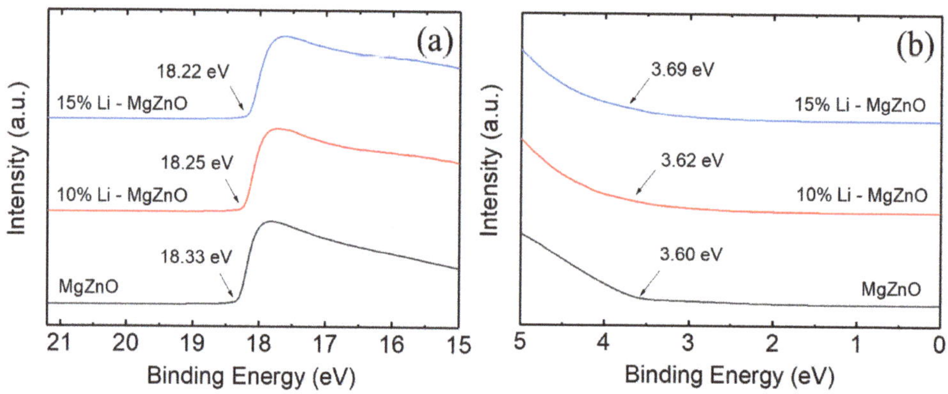

Figure 4. UPS spectra of (**a**) the secondary cutoff region and (**b**) the valence band maximum region for the Li-doped MgZnO nanoparticles.

Figure 5 shows the photoluminescence (PL) spectra of MgZnO, 10% and 15% Li-doped MgZnO nanoparticles. As shown in Figure 3, since the band gap of Li-doped MgZnO nanoparticles exceeds 3.5 eV, the visible emission is presumed to originate mainly from surface defects [57]. MgZnO nanoparticles exhibit a broad emission peak around 523 nm. With Li doping, the emission peak shifts toward the shorter wavelength region, appearing at 507 nm and 494 nm for 10% and 15% Li-doped MgZnO nanoparticles, respectively. The visible emission wavelength of MgZnO nanoparticles depends on their band gap [20,58]. As shown in Figure 3, the band gap increases with Li doping, resulting in a blue shift of the defect related emission wavelength. This occurs because the increased band gap shifts the conduction and valence band edges while the absolute position of the defect level remains constant, leading to a higher energy difference between defect level and the conduction band minimum, as reported by Pan et al. [20]. In addition, the PL intensity decreases with Li doping. Since PL intensity depends on the amount of surface defects, the lower PL intensity in Li-doped MgZnO nanoparticles indicates the reduced surface defects [30,57]. During the formation of nanoparticles, the Li ions may be attached to the surface and passivate the surface defects. A similar phenomenon has also been observed when Mg is doped into ZnO nanoparticles [44].

Figure 6 shows the AFM images of the thin films coated with MgZnO, 10% and 15% Li-doped MgZnO nanoparticles. The root mean square (RMS) surface roughness of the MgZnO nanoparticle film is 2.6 nm, while the 10% and 15% Li-doped MgZnO films exhibit RMS surface roughness of 2.4 nm and 2.8 nm, respectively. On the other hand, the particle sizes shown in Figure 2 are 3.6, 3.0, and 2.7 nm for the undoped, 10% and 15% Li-doped MgZnO nanoparticles, resulting in the discrepancy between the particle size observed by TEM and the surface roughness measured by AFM. This apparent inconsistency can be attributed to the agglomeration of nanoparticles [59]. As the particle size decreases, the surface energy of nanoparticles increases significantly, which promotes agglomeration during film formation. Especially for the smallest particles of 2.7 nm, the reduced stability leads to local agglomeration, resulting in uneven surface morphology and increased surface roughness, despite the smaller individual particle size.

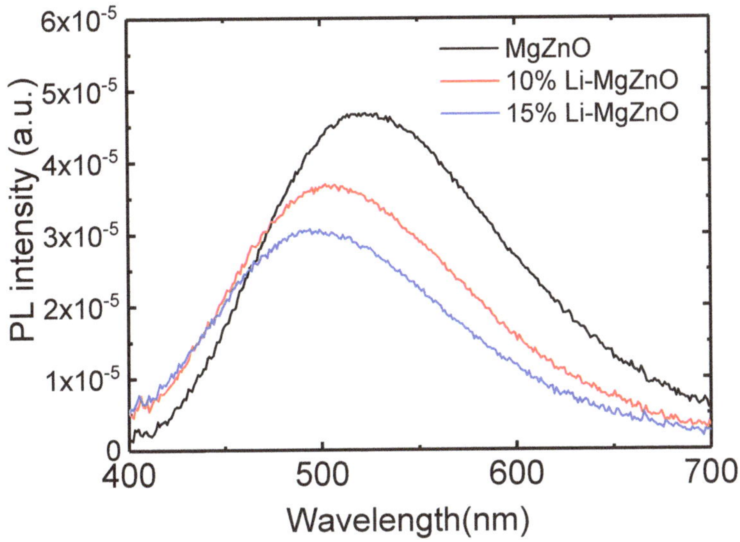

Figure 5. PL spectra of the Li-doped ZnO nanoparticles.

Figure 6. AFM images of MgZnO, 10% Li-doped MgZnO, and 15% Li-doped MgZnO nanoparticle films.

To investigate the electrical conduction properties of Li-doped MgZnO nanoparticles, ITO/Li-doped MgZnO nanoparticles (80 nm)/Al devices were prepared. Figure 7 shows the current density–voltage curves of the devices. Compared to undoped MgZnO nanoparticles, the Li-doped MgZnO nanoparticles exhibit reduced current transport. For example, MgZnO nanoparticle device shows a current density of 544 mA/cm^2 at 5 V, whereas the 10% and 15% Li-doped MgZnO nanoparticles exhibit current densities of 321 and 79 mA/cm^2, respectively, at the same voltage. The electron mobility of the Li-doped MgZnO nanoparticles was estimated in the space charge limited current region using the equation, $J = (9/8)\mu\varepsilon_o\varepsilon_r(V^2/L^3)$, where J is the current density, μ is the electron mobility, ε_o is the dielectric constant of free space, ε_r is the relative dielectric constant of the nanoparticle film, V is the applied voltage, and L is the thickness of the nanoparticle film [60]. The mobility of the undoped MgZnO nanoparticle device was approximately 10^{-5} cm^2/Vs, while the mobilities of the 10% and 15% Li-doped MgZnO nanoparticle devices were 5×10^{-6} and 9×10^{-7} cm^2/Vs, respectively, indicating a decrease in mobility with increasing Li concentration. This reduction in mobility is presumed to be related to the decreased particle size and impurity effects induced by Li doping. As the particle size decreases with Li doping as shown in Figure 2, the electron scattering at the particle surfaces increases. Additionally, impurity scattering due to Li dopants also contributes to the decrease in mobility.

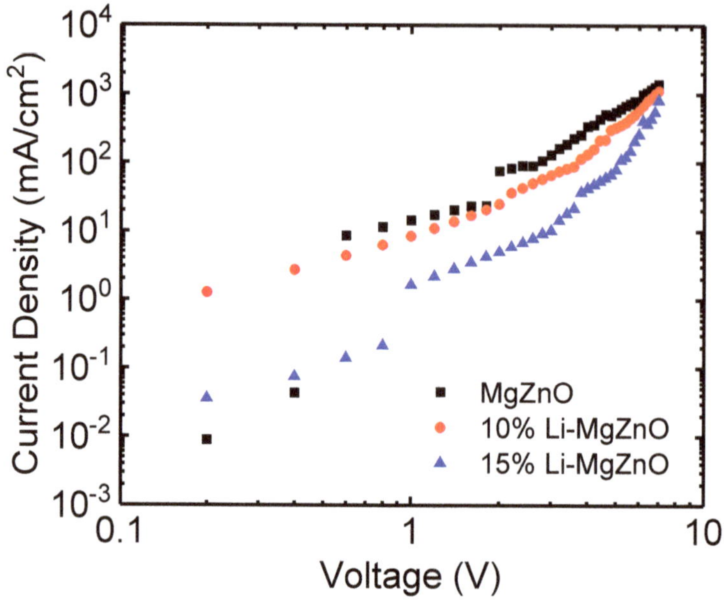

Figure 7. Current density–voltage curves of the Li-doped MgZnO nanoparticle films.

Inverted OLEDs were fabricated using Li-doped MgZnO nanoparticles. The device structure is ITO/Li-doped MgZnO nanoparticles (10 nm)/TPBi (25 nm)/CBP:Ir(ppy)$_3$ (20 nm, 5 wt%)/CBP (30 nm)/MoO_3 (10 nm)/Al. Figure 8 shows the device structure and energy diagram of the inverted devices with Li-doped MgZnO nanoparticles. Figure 9 shows the electroluminescence spectra measured at various voltages for the inverted devices fabricated using MgZnO, 10% and 15% Li-doped MgZnO nanoparticles. The devices exhibit a strong emission at 515 nm, which originates from the triplet excited state of Ir(ppy)$_3$ [61]. The emission from Ir(ppy)$_3$ remains unchanged with varying voltage. In addition, a weak emission peak can be observed around 400–410 nm. This peak corresponds to the emission from CBP and increases with increasing voltage [62]. The CBP emission peak indicates electron leakage from the emission layer into the CBP hole transport layer, implying that charge balance is dominated by electron conduction. Furthermore, this result suggests that the recombination zone is located near the emission layer adjacent to the CBP hole transport layer. As the voltage increases, the CBP emission peak becomes more pronounced, indicating increased electron leakage into CBP layer. Moreover, the devices based on 10% and 15% Li-doped MgZnO nanoparticles exhibit weaker CBP emission intensity compared to the device with undoped MgZnO nanoparticles. This suggests that charge balance is improved in the devices using Li-doped MgZnO nanoparticles.

Figure 10 shows the current density–voltage–luminance and external quantum efficiency (EQE) curves of the inverted OLEDs using Li-doped MgZnO nanoparticles. It can be observed that the driving voltage of the Li-doped MgZnO nanoparticle devices is higher. For example, in the case of the undoped MgZnO device, the driving voltage for a current density of 20 mA/cm^2 is 12.4 V. In contrast, the devices with 10% and 15% Li-doped ZnO nanoparticles have driving voltages of 13.8 V and 14.4 V, respectively, to achieve the same current density. This increase in driving voltage is consistent with the suppression of electron transport in Li-doped MgZnO nanoparticle, as shown in Figure 7. The turn-on voltage for obtaining a luminance of 1 cd/m^2 also slightly increases with Li doping. The turn-on voltages for MgZnO, 10%, and 15% Li-doped MgZnO nanoparticle devices are 6.4, 6.8, and 6.8 V, respectively. The turn-on voltages are quite high compared to the other inverted OLEDs [13–16,63]. For example, Hwang et al. reported turn-on voltages of 3.4–3.8 V in the inverted devices with ZnO nanoparticles [63]. The high turn-on voltages in the Li-doped MgZnO devices may be attributed to the low electron mobilities of both the Li-doped MgZnO nanoparticles and the TPBi electron transport layer. The electron mobility of ZnO nanoparticles has been reported to be approximately 6×10^{-4} to 5×10^{-3} cm^2/Vs [20]. On the other hand, the electron mobility of Li-doped MgZnO nanoparticles is estimated to be in the range of 9×10^{-7} to 10^{-5} cm^2/Vs, based on Figure 7. In addition, the electron mobility of TPBi ranges from 5.6×10^{-8} to 2.1×10^{-5} cm^2/Vs [64]. These low mobilities lead to a significant voltage drop across the Li-doped MgZnO and TPBi layers, resulting in high turn-on voltages. The voltage required for high luminance also increases with Li doping. For example, in the case of MgZnO, the voltage for 1000 cd/m^2 luminance is 9.6 V, whereas for 10% and 15% Li-doped MgZnO devices, it increases to 11.8 V and 12.4 V, respectively. Figure 10b shows the EQE curves for the Li-doped MgZnO nanoparticle devices. The maximum EQE of the Li-doped MgZnO devices is higher than that of the undoped devices. The MgZnO device shows a maximum EQE of 17.5% at 296 cd/m^2, while the 10% and 15% Li-doped MgZnO nanoparticle devices show maximum EQEs of 18.5% at 313 cd/m^2 and 21.7% at 683 cd/m^2, respectively, indicating better charge balance in the Li-doped MgZnO nanoparticle devices.

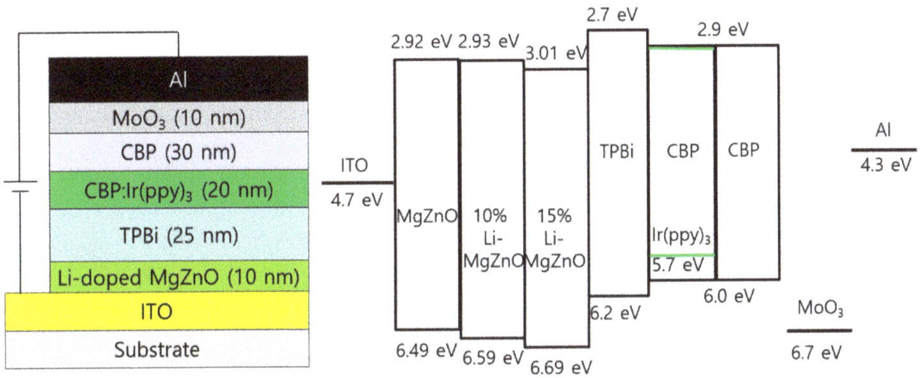

Figure 8. Device structure and energy diagram of the inverted OLEDs with Li-doped MgZnO nanoparticles.

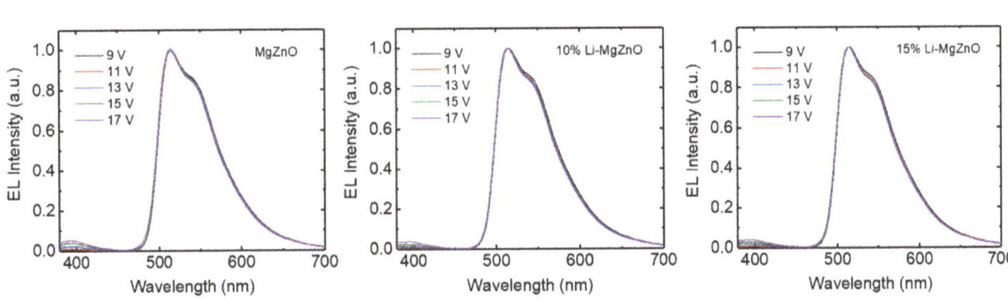

Figure 9. EL spectra at various voltages for the inverted OLEDs with Li-doped MgZnO nanoparticles.

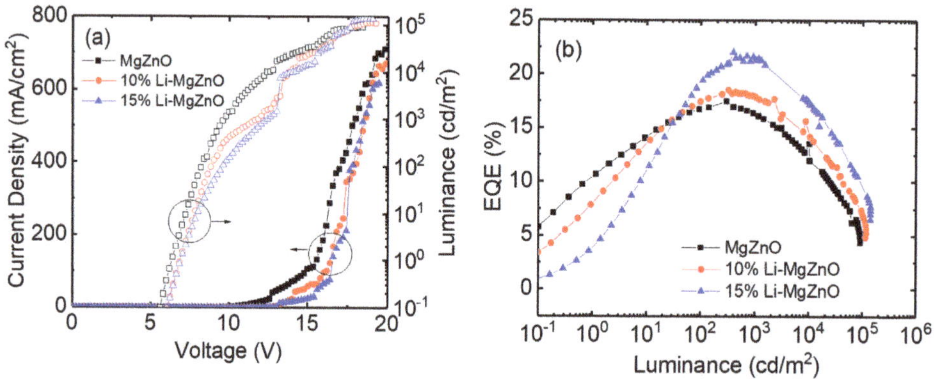

Figure 10. (**a**) Current density—voltage—luminance and (**b**) EQE curves for the inverted OLEDs with Li-doped MgZnO nanoparticles.

4. Conclusions

Li-doped MgZnO nanoparticles with a hexagonal wurtzite structure were synthesized using LiCl, magnesium acetate tetrahydrate, zinc acetate dihydrate, and TMAH for the electron injection layer in the inverted OLEDs. Li doping inhibits particle growth, resulting in smaller particle sizes. The average particle size of MgZnO nanoparticles was 3.6 nm, whereas those of 10% and 15% Li-doped MgZnO nanoparticles were 3.0 and 2.7 nm, respectively. The band gap of Li-doped MgZnO nanoparticles was 3.57 eV, whereas the band gap increased due to the enhanced quantum confinement effects resulting from the reduced particle size, reaching 3.68 eV for the 15% Li-doped ZnO nanoparticles. Li doping also raised both the VBM and CBM energy levels. The VBM energy of MgZnO nanoparticles was 6.49 eV, while those of 10% and 15% Li-doped MgZnO nanoparticles were 6.59 and 6.69 eV, respectively. The CBM energy of MgZnO nanoparticles was 2.92 eV and increased to 3.01 eV with 15% Li doping. The increase in CBM energy due to Li doping indicates an increase in energy barrier for the electron injection from the nanoparticle layer into the TPBi electron transport layer. The Li-doped MgZnO nanoparticles exhibited a visible emission due to the surface defects, and the emission spectrum shifted toward shorter wavelength region with Li doping, accompanied by a decrease in emission intensity. The electron mobility of MgZnO nanoparticles was about 10^{-5} cm^2/Vs, which decreased with Li doping to 5×10^{-6} and 9×10^{-7} cm^2/Vs for the 10% and 15% Li-doped MgZnO nanoparticles, respectively. The inverted phosphorescent OLEDs using Li-doped MgZnO nanoparticles exhibited a suppressed electron conduction, resulting in better charge balance and higher external quantum efficiency compared to the devices using undoped MgZnO nanoparticles. The maximum EQE reached 21.7% for the device with 15% Li-doped MgZnO nanoparticles.

Author Contributions: Conceptualization, D.-G.M.; methodology, H.-J.Y., G.-E.K., C.-J.P., S.-B.L. and S.-Y.K.; validation, D.-G.M.; data curation, H.-J.Y. and G.-E.K.; writing—original draft preparation, H.-J.Y.; writing—review and editing, D.-G.M.; supervision, D.-G.M.; project administration, D.-G.M.; funding acquisition, D.-G.M. All authors have read and agreed to the published version of the manuscript.

Funding: This work was supported by the Soonchunhyang University Research Fund and Korea Institute for Advancement of Technology (KIAT) grant funded by the Korea government (MOTIE) (RS-2025-02263458, HRD Program for Industrial Innovation), and the Super-Gap R&D Project (RS-2024-00420455) of MOTIE/KEIT.

Data Availability Statement: Data are contained within the article.

Conflicts of Interest: The authors declare no conflicts of interest.

References

1. Tang, C.W.; VanSlyke, S.A. Organic electroluminescent diodes. *Appl. Phys. Lett.* **1987**, *51*, 913–915. [CrossRef]
2. Baldo, M.A.; O'Brien, D.F.; You, Y.; Shoustikov, A.; Sibley, S.; Thompson, M.E.; Forrest, S.R. Highly efficient phosphorescent emission from organic electroluminescent devices. *Nature* **1998**, *395*, 151–154. [CrossRef]
3. Hong, G.; Gan, X.; Leonhardt, C.; Zhang, Z.; Seibert, J.; Busch, J.M.; Bräse, S. A brief history of OLEDs-emitter development and industry milestones. *Adv. Mater.* **2021**, *33*, 2005630. [CrossRef]
4. Bauri, J.; Choudhary, R.B.; Mandal, G. Recent advances in efficient emissive materials-based OLED applications: A review. *J. Mater. Sci.* **2021**, *56*, 18837–18866. [CrossRef]
5. Huseynova, G.; Lee, J.H.; Kim, Y.H.; Lee, J. Transparent organic light-emitting diodes: Advances, prospects, and challenges. *Adv. Opt. Mater.* **2021**, *9*, 2002040. [CrossRef]
6. Chang, S.; Koo, J.H.; Yoo, J.; Kim, M.S.; Choi, M.K.; Kim, D.H. Flexible and stretchable light-emitting diodes and photodetectors for human-centric optoelectronics. *Chem. Rev.* **2024**, *124*, 768–859. [CrossRef]
7. Ma, R.Q.; Hewitt, R.; Rajan, K.; Silvernail, J.; Urbanik, K.; Brown, J.J. Flexible active-matrix OLED displays: Challenges and progress. *J. Soc. Inf. Disp.* **2008**, *16*, 169–175. [CrossRef]
8. Su, Y.; Geng, D.; Chen, Q.; Ji, H.; Li, M.; Shang, G.; Liu, L.; Chuai, X.; Huang, S.; Lu, N.; et al. Novel TFT-based emission driver in high performance AMOLED display applications. *Org. Electron.* **2021**, *93*, 106160. [CrossRef]
9. Park, J.S.; Maeng, W.J.; Kim, H.S.; Park, J.S. Review of recent developments in amorphous oxide semiconductor thin-film transistor devices. *Thin Solid Films* **2012**, *520*, 1679–1693. [CrossRef]
10. Zhu, H.; Shin, E.S.; Liu, A.; Ji, D.; Xu, Y.; Noh, Y.Y. Printable semiconductors for backplane TFTs of flexible OLED displays. *Adv. Func. Mater.* **2020**, *30*, 1904588. [CrossRef]
11. Shi, J.; Zhang, J.; Yang, L.; Qu, M.; Qi, D.C.; Zhang, K.H.L. Wide bandgap oxide semiconductors: From materials physics to optoelectronic devices. *Adv. Mater.* **2021**, *33*, 2006230. [CrossRef] [PubMed]
12. Chu, T.Y.; Chen, J.F.; Chen, S.Y.; Chen, C.J.; Chen, C.H. Highly efficient and stable inverted bottom-emission organic light emitting devices. *Appl. Phys. Lett.* **2006**, *89*, 053503. [CrossRef]
13. Wang, J.; Zhang, M.; Zhang, Y.; Fu, J.; Qin, Y.; Li, R.; Chen, Y.; Lai, W.; Zhang, X.; Huang, W. Efficient inverted organic light-emitting devices using a charge-generation unit as electron-injection layers. *Org. Electron.* **2021**, *96*, 106202. [CrossRef]
14. Yang, H.I.; Jeong, S.H.; Cho, S.M.; Lampande, R.; Lee, K.M.; Hong, J.A.; Choi, J.W.; Kim, B.S.; Park, Y.; Pode, R.; et al. Efficient cathode contacts through Ag-doping in multifunctional strong nucleophilic electron transport layer for high performance inverted OLEDs. *Org. Electron.* **2021**, *89*, 106031. [CrossRef]
15. Shi, G.; Zhang, X.; Wan, M.; Wang, S.; Lian, H.; Xu, R.; Zhu, W. High-performance inverted organic light-emitting didoes with extremely low efficiency roll-off using solution-processed ZnS quantum dots as the electron injection layer. *RSC Adv.* **2019**, *9*, 6042–6047. [CrossRef]
16. Chang, C.H.; Hsu, M.K.; Wu, S.W.; Chen, M.H.; Lin, H.H.; Li, C.S.; Pi, T.W.; Chang, H.H.; Chen, N.P. Using lithium carbonate-based electron injection structures in high-performance inverted organic light-emitting diodes. *Phys. Chem. Chem. Phys.* **2015**, *17*, 13123–13128. [CrossRef] [PubMed]
17. Tang, X.; Ding, L.; Sun, Y.Q.; Xie, Y.M.; Deng, Y.L.; Wang, Z.K.; Lia, L.S. Inverted and large flexible organic light-emitting diodes with low operation voltage. *J. Mater. Chem. C* **2015**, *3*, 12399–12402. [CrossRef]
18. Chen, J.; Shi, C.; Fu, Q.; Zhao, F.; Hu, Y.; Feng, Y.; Ma, D. Solution-processable small molecules as efficient universal bipolar host for blue, green and red phosphorescent inverted OLEDs. *J. Mater. Chem.* **2012**, *22*, 5164–5170. [CrossRef]
19. Lee, H.; Kang, C.M.; Park, M.; Kwak, J.; Lee, C. Improved efficiency of inverted organic light-emitting diodes using tin dioxide nanoparticles as an electron injection layer. *ACS Appl. Mater. Interfaces* **2013**, *5*, 1977–1981. [CrossRef]
20. Pan, J.; Chen, J.; Huang, Q.; Khan, Q.; Liu, X.; Tao, Z.; Zhang, Z.; Lei, W.; Nathan, A. Size tunable ZnO nanoparticles to enhance electron injection in solution processed QLEDs. *ACS Photonics* **2016**, *3*, 215–222. [CrossRef]
21. Kirkwood, N.; Singh, B.; Mulvaney, P. Enhancing quantum dot LED efficiency by tuning mobility in the ZnO electron transport layer. *Adv. Mater. Interfaces* **2016**, *3*, 1600868. [CrossRef]
22. Sessolo, M.; Bolink, H.J. Hybrid organic-inorganic light-emitting diodes. *Adv. Mater.* **2011**, *23*, 1829–1845. [CrossRef] [PubMed]
23. Youn, H.; Yang, M. Solution processed polymer light-emitting diodes utilizing a ZnO/organic ionic interlayer with Al cathode. *Appl. Phys. Lett.* **2010**, *97*, 243302. [CrossRef]
24. Qian, L.; Zheng, Y.; Choudhury, K.R.; Bera, D.; So, F.; Xue, J.; Holloway, P.H. Electroluminescence from light-emitting polymer/ZnO nanoparticle heterojunction at sub-bandgap voltages. *Nano Today* **2010**, *5*, 384–389. [CrossRef]

25. Chiba, T.; Pu, Y.J.; Hirasawa, M.; Masuhara, A.; Sasabe, H.; Kido, J. Solution-processed inorganic-organic hybrid electron injection layer for polymer light-emitting devices. *ACS Appl. Mater. Interfaces* **2012**, *4*, 6104–6108. [CrossRef]
26. Lee, B.R.; Jung, E.D.; Park, J.S.; Nam, Y.S.; Min, S.H.; Kim, B.S.; Lee, K.M.; Jeong, J.R.; Friend, R.H.; Kim, J.S.; et al. Highly efficient inverted polymer light-emitting diodes using surface modifications of ZnO layer. *Nat. Commun.* **2014**, *5*, 4840. [CrossRef]
27. Pu, Y.J.; Morishita, N.; Chiba, T.; Ohisa, S.; Igarashi, M.; Masuhara, A.; Kido, J. Efficient electron injection by size- and shape-controlled zinc oxide nanoparticles in organic light-emitting devices. *ACS Appl. Mater. Interfaces* **2015**, *7*, 25373–25377. [CrossRef]
28. Murat, Y.; Langer, E.; Barnes, J.P.; Laurent, J.Y.; Wantz, G.; Hirsch, L.; Maindron, T. Bright and efficient inverted organic light-emitting diodes with improved solution-processed electron-transport interlayers. *Org. Electron.* **2017**, *48*, 377–381. [CrossRef]
29. Kaçar, R.; Mucur, S.P.; Yildiz, F.; Dabak, S.; Tekin, E. Solution processed ternary blend nano-composite charge regulation layer to enhance inverted OLED performances. *Appl. Phys. Lett.* **2018**, *112*, 163302. [CrossRef]
30. Khairnar, N.; Kwon, H.; Park, S.; Lee, H.; Park, J. Tailoring the size and shape of ZnO nanoparticles for enhanced performance of OLED device. *Nanomaterials* **2023**, *13*, 2816. [CrossRef]
31. Lu, L.P.; Kabra, D.; Friend, R.H. Barium hydroxide as an interlayer between zinc oxide and a luminescent conjugated polymer for light-emitting didoes. *Adv. Funct. Mater.* **2012**, *22*, 4165–4171. [CrossRef]
32. Hőfle, S.; Schienle, A.; Bernhard, C.; Bruns, M.; Lemmer, U.; Colsmann, A. Solution processed, white emitting tandem organic light-emitting diodes with inverted device architecture. *Adv. Mater.* **2014**, *26*, 5155–5159. [CrossRef] [PubMed]
33. Zhang, J.; Zhang, X.; Feng, H.; Yu, Z.; Zhang, J.; Liu, S.; Zhang, L.; Xie, W. An efficient and stable hybrid organic light-emitting device based on an inorganic metal oxide hole transport layer and an electron transport layer. *J. Mater. Chem. C* **2019**, *7*, 1991–1998. [CrossRef]
34. Zhou, L.; Xiang, H.Y.; Zhu, Y.F.; Ou, Q.D.; Wang, Q.K.; Hu, R.; Huang, X.B.; Tang, J.X. Multifunctional silver nanoparticle interlayer-modified ZnO as the electron-injection layer for efficient inverted organic light-emitting diodes. *ACS Appl. Mater. Interfaces* **2019**, *11*, 9251–9258. [CrossRef]
35. Chiba, T.; Pu, Y.J.; Kido, J. Solution-processable electron injection materials for organic light-emitting devices. *J. Mater. Chem. C* **2015**, *3*, 11567–11576. [CrossRef]
36. Eun, Y.B.; Jang, G.P.; Yang, J.H.; Kim, S.Y.; Chae, Y.B.; Ha, M.Y.; Moon, D.G.; Kim, C.K. Performance improvement of quantum dot light-emitting diodes using a ZnMgO electron transport layer with a core/shell structure. *Materials* **2023**, *16*, 600. [CrossRef]
37. Alexandrov, A.; Zvaigzne, M.; Lypenko, D.; Nabiev, I.; Samokhvalov, P. Al-, Ga-, Mg-, or Li-doped zinc oxide nanoparticles as electron transport layers for quantum dot light-emitting diodes. *Sci. Rep.* **2020**, *10*, 7496. [CrossRef] [PubMed]
38. Kim, H.M.; Cho, S.; Kim, J.; Shin, H.; Jang, J. Li and Mg co-doped zinc oxide electron transporting layer for highly efficient quantum dot light-emitting diodes. *ACS Appl. Mater. Interfaces* **2018**, *10*, 24028–24036. [CrossRef]
39. Li, Y.; Dai, X.; Chen, D.; Ye, Y.; Gao, Y.; Peng, X.; Jin, Y. Inverted quantum dot light-emitting diodes with conductive interlayers of zirconium acetylacetonate. *J. Mater. Chem. C* **2019**, *7*, 3154–3159. [CrossRef]
40. Manzhi, P.; Alam, M.B.; Kumari, R.; Krishna, R.; Singh, R.K.; Srivastava, R.; Sinha, O.P. Li-doped ZnO nanostructures for the organic light emitting diode application. *Vacuum* **2017**, *146*, 462–467. [CrossRef]
41. Manzhi, P.; Kumari, R.; Alam, M.B.; Umapathy, G.R.; Krishna, R.; Ojha, S.; Srivastava, R.; Sinha, O.P. Mg-doped ZnO nanostructures for efficient organic light emitting diode. *Vacuum* **2019**, *166*, 370–376. [CrossRef]
42. Kim, J.H.; Han, C.Y.; Lee, K.H.; An, K.S.; Song, W.; Kim, J.; Oh, M.S.; Do, Y.R.; Yang, H. Performance improvement of quantum dot-light-emitting diodes enabled by an alloyed ZnMgO nanoparticle electron transport layer. *Chem. Mater.* **2015**, *27*, 197–204. [CrossRef]
43. Wang, S.; Guo, Y.; Feng, D.; Chen, L.; Fang, Y.; Shen, H.; Du, Z. Bandgap tunable $Zn_{1-x}Mg_xO$ thin films as electron transport layers for high performance quantum dot light-emitting diodes. *J. Mater. Chem. C* **2017**, *5*, 4724–4730. [CrossRef]
44. Chen, L.; Zhang, Y.; Kun, Y.; Tuo, K.; Shang, J.; Du, W.; Qi, H.; Liu, S. The $Zn_{1-x}Mg_xO$ electron transport layer for charge balance in high-brightness inverted quantum-dot light-emitting diodes. *J. Mater. Sci. Mater. Electron.* **2024**, *35*, 754. [CrossRef]
45. Wang, L.; Lin, J.; Liu, X.; Cao, S.; Wang, Y.; Zhao, J.; Zou, B. Mg-doped ZnO nanoparticle films as the interlayer between the ZnO electron transport layer and InP quantum dot layer for light-emitting diodes. *J. Phys. Chem. C* **2020**, *124*, 8758–8765. [CrossRef]
46. Pacholski, C.; Kornowski, A.; Weller, H. Self-assembly of ZnO: From nanodots to nanorods. *Angew. Chem. Int. Ed.* **2002**, *41*, 1188–1191. [CrossRef]
47. Moyen, E.; Kim, J.H.; Kim, J.; Jang, J. ZnO nanoparticles for quantum-dot-based light-emitting diodes. *ACS Appl. Nano Mater.* **2020**, *3*, 5203–5211. [CrossRef]
48. Ahmoum, H.; Boughrara, M.; Su'ait, M.S.; Kerouad, M. Effect of position and concentration of Li on ZnO physical properties: Density functional investigation. *Chem. Phys. Lett.* **2019**, *719*, 45–53. [CrossRef]
49. Vidya, R.; Ravindran, P.; Fjellvåg, H. Ab-initio studies on Li-doping, Li-pairs, and complexes between Li and intrinsic defects in ZnO. *J. Appl. Phys.* **2012**, *111*, 123713. [CrossRef]
50. Hirai, T.; Harada, Y.; Hashimoto, S.; Itoh, T.; Ohno, N. Luminescence of excitons in mesoscopic ZnO particles. *J. Lumin.* **2005**, *112*, 196–199. [CrossRef]

51. Brus, L. Electronic wave functions in semiconductor clusters: Experimental and theory. *J. Phys. Chem.* **1986**, *90*, 2555–2560. [CrossRef]
52. Viswanatha, R.; Sapra, S.; Satpati, B.; Satyam, P.V.; Dev, B.N.; Sarma, D.D. Understanding the quantum size effects in ZnO nanocrystals. *J. Mater. Chem.* **2004**, *14*, 661–668. [CrossRef]
53. Gil, B.; Kavokin, A.V. Giant exciton-light coupling in ZnO quantum dots. *Appl. Phys. Lett.* **2002**, *81*, 748–750. [CrossRef]
54. Igityan, A.; Aghamalyan, N.; Petrosyan, S.; Badalyan, G.; Kafadaryan, Y. Resistive switching in Li-doped ZnO films. *Phys. Status Solidi A* **2018**, *215*, 1700353. [CrossRef]
55. Aida, M.S.; Hjiri, M. Temperature-dependent photoluminescence of Li-doped ZnO. *J. Mater. Sci. Mater. Electron.* **2020**, *31*, 10521–10530. [CrossRef]
56. Cao, S.; Zheng, J.; Zhao, J.; Yang, Z.; Li, C.; Guan, X.; Yang, W.; Shang, M.; Wu, T. Enhancing the performance of quantum dot light-emitting diodes using room-temperature-processed Ga-doped ZnO nanoparticles as the electron transport layer. *ACS Appl. Mater. Interfaces* **2017**, *9*, 15605–15614. [CrossRef] [PubMed]
57. Zhang, L.; Yin, L.; Wang, C.; Lun, N.; Qi, Y.; Xiang, D. Origin of visible photoluminescence of ZnO quantum dots: Defect-dependent and size-dependent. *J. Phys. Chem. C* **2010**, *114*, 9651–9658. [CrossRef]
58. Li, N.; Li, T.; Li, L.; Li, J. Efficient and stable OLEDs with inverted device structure utilizing solution-processed ZnO-based electron injection layer. *Adv. Opt. Mater.* **2023**, *11*, 2300467. [CrossRef]
59. Lim, S.J.; Kim, H.; Hwang, H.A.; Park, H.J.; Moon, D.G. Effect of oxidizing agent on the synthesis of ZnO nanoparticles for inverted phosphorescent organic light-emitting devices without multiple interlayers. *Nanomaterials* **2024**, *14*, 622. [CrossRef]
60. Yuan, K.; Chen, L.; Li, F.; Chen, Y. Nanostructured hybrid ZnO@CdS nanowalls grown in situ for inverted polymer solar cells. *J. Mater. Chem. C* **2014**, *2*, 1018–1027. [CrossRef]
61. Hofbeck, T.; Yersin, H. The triplet state of fac-Ir(ppy)$_3$. *Inorg. Chem.* **2010**, *49*, 9290–9299. [CrossRef] [PubMed]
62. Jou, J.H.; Sun, M.C.; Chou, H.H.; Li, C.H. Efficient pure-white organic light-emitting diodes with a solution-processed, binary-host employing single emission layer. *Appl. Phys. Lett.* **2006**, *88*, 141101. [CrossRef]
63. Hwang, H.A.; Park, H.J.; Moon, D.G. Highly efficient inverted phosphorescent organic light-emitting devices with ZnO nanoparticles electron injection layer. *Synth. Met.* **2022**, *287*, 117078. [CrossRef]
64. Wang, Y.; Li, B.; Jiang, C.; Fang, Y.; Bai, P.; Wang, Y. Study on electron transport characterization in TPBi thin films and OLED application. *J. Phys. Chem. C* **2021**, *125*, 16753–16758. [CrossRef]

micromachines

MDPI

Article

TCAD Simulation Study of Electrical Performance of a Novel High-Purity Germanium Drift Detector

Mingyang Wang [1], Zheng Li [1,2,*], Bo Xiong [1] and Yongguang Xiao [1,*]

[1] School of Materials Science and Engineering, Xiangtan University, Xiangtan 411105, China; 202131550147@smail.xtu.edu.cn (M.W.); boxiong65190@163.com (B.X.)
[2] College of Integrated Circuits, Ludong University, Yantai 264025, China
* Correspondence: 3636@ldu.edu.cn (Z.L.); ygxiao@xtu.edu.cn (Y.X.)

Abstract: High-purity germanium (HPGe) detectors occupy a prominent position in fields such as radiation detection and aerospace because of their excellent energy resolution and wide detection range. To achieve a broader detection range, conventional HPGe detectors often need to be expanded to cubic-centimeter-scale volumes. However, this increase in volume leads to a large detector area, which in turn increases the detector capacitance, affecting the detector's noise level and performance. To address this issue, this study proposes a novel high-purity germanium drift detector (HPGeDD). The design features a small-area central collecting cathode surrounded by concentric anode rings, with a resistive chain interposed between the anode rings to achieve self-dividing voltage. This design ensures that the detector's capacitance is only related to the area of the central collecting cathode, independent of the overall active area, thus achieving a balance between a small capacitance and large active area. Electrical performance simulations of the novel detector were conducted using the semiconductor simulation software Sentaurus TCAD (P-2019.03). The results show a smooth electric potential distribution within the detector, forming a lateral electric field, as well as a lateral hole drift channel precisely directed toward the collecting cathode. Furthermore, simulations of heavy ion incidence were performed to investigate the detector's carrier collection characteristics. The simulation results demonstrate that the HPGeDD exhibits advantages such as fast signal response and short collection time. The design proposal presented in this study offers a new solution to the problem of excessive capacitance in conventional HPGe detectors, expands their application scope, and provides theoretical guidance for subsequent improvements, optimizations, and practical manufacturing.

Keywords: high-purity germanium drift detector; TCAD simulation of electrical performance; heavy ion incidence

1. Introduction

Since the mid-1970s, with the continuous improvement in material purification technologies, particularly the rapid development of zone melting techniques, the impurity concentration in germanium crystals has been significantly reduced [1]. As a type of semiconductor detector, HPGe detectors demonstrate distinct advantages over silicon-based detectors. This superiority primarily stems from the fact that germanium possesses a higher atomic number, a lower average ionization energy, and a higher charge carrier mobility. These physical characteristics enable HPGe detectors to generate larger signals when detecting the same amount of radiation energy. Consequently, compared to silicon-based detectors, HPGe detectors theoretically exhibit wider energy responses, faster signal response times, higher collection rates, and stronger signal outputs. Additionally, HPGe detectors have the capability to be stored at room temperature and perform optimally when operated in a liquid-nitrogen-cooled environment [2,3].

HPGe detectors exhibit practical application value in the detection of heavy ions and other radiation particles [4,5]. Due to their extremely low internal radioactivity levels and high energy resolution, HPGe detectors have become key devices in the field of dark matter detection. When external radiation particles enter the sensitive volume of the detector, they induce the generation of electron–hole pairs. Under the influence of the internal electric field, these charge carriers are efficiently collected by the collecting electrodes, converting the radiation signal into an electrical signal. For example, the CDEX research team, composed of Chinese scientists, utilizes a ton-scale point-contact electrode HPGe detector array system to conduct direct dark matter detection experiments at the China Jinping Underground Laboratory [6]. The aim is to detect light dark matter particles with masses less than 10 GeV/c^2, obtaining information such as the energy and position of the incident particles.

On the other hand, HPGe detectors hold potential application value in the field of X-ray pulsar navigation. Pulsars, as the most precise natural astronomical clocks, exhibit extremely stable rotation periods, with typical values ranging from milliseconds to seconds with a time precision better than 10^{-7} s [7–11]. These stable rotation periods make pulsars ideal candidates for various applications, including X-ray pulsar navigation, which is a novel autonomous navigation technology for spacecraft, utilizing pulsars as navigation beacons to provide autonomous navigation services for spacecraft in near-Earth orbits, deep space, and interstellar space [12–17]. Semiconductor detectors are the core components of X-ray pulsar navigation systems. When X-rays enter the depletion region of a semiconductor detector, they interact with material atoms to generate electron–hole pairs, which are then absorbed by the collecting electrodes under the influence of an electric field, thereby converting the X-ray signals emitted by pulsars into electrical signals.

Currently, silicon-based detectors have been applied in X-ray pulsar navigation systems [18]. Due to their superior material properties compared to silicon-based detectors, HPGe detectors offer performance advantages, such as further extension of the detection range and improved energy resolution, making them promising candidates for future development. With the further advancement of space cooling technologies, an aerospace-grade pulse tube cooler, successfully developed by a research team at the Institute of Physics and Chemistry of the Chinese Academy of Sciences, has been applied in the aerospace field, providing a reliable low-temperature liquid nitrogen environment. This has made it possible to realize the application of HPGe detectors in the field of X-ray pulsar navigation.

The application of HPGe detectors in the fields of radiation particles and X-ray detection is fundamentally based on the photoelectric effect of germanium crystals. When an HPGe detector is subjected to external radiation, the germanium crystal within the depletion region undergoes ionization, generating electron–hole pairs. Under the influence of the internal electric field, these charge carriers drift toward the corresponding electrodes and are ultimately collected by the detector's electrodes. The resulting electrical signals are then amplified and processed for analyzing the properties of the radiation energy.

This has greatly promoted the remarkable advancement in HPGe detectors. HPGe detectors are primarily classified into two types: planar and coaxial. Planar detectors, distinguished by their compact structure, are particularly suitable for applications in confined spaces or specific scenarios, and they demonstrate excellent performance in measuring medium-energy to high-energy charged particle, nuclear radiation, as well as low-energy gamma rays [19]. Coaxial detectors, on the other hand, leverage their advantages of long axial detection length and large sensitive volume, making them an ideal choice for measuring high-energy rays with strong penetration capabilities [20].

HPGe detectors typically feature an ultra-large sensitive volume, aimed at expanding the detection range, capturing more detection signals, enhancing the signal collection capability, improving the detection efficiency, and, thus, more accurately reflecting the characteristics of the object being measured [21]. However, the increased thickness of the large-sized sensitive volume also leads to an increase in capacitance, which adversely affects the detector's noise level and overall performance. Attempting to reduce capacitance by decreasing size often comes at the expense of the detector's active area for detection. Therefore, we are faced with the challenge of effectively reducing capacitance while ensuring that the detector's active area is not compromised.

To address the challenge of expanding the detection area of HPGe detectors while maintaining low capacitance, this study proposes an innovative detector design. The HPGe detector is designed as a large-area drift detector structure with a small-area central collecting cathode surrounded by 20 concentric anode rings, with a resistor chain capable of automatic voltage division added between the anode rings. Based on the structural characteristics of semiconductor drift detectors [22], this design ensures that the output capacitance of the detector is solely dependent on the area of the central collecting cathode, rather than the entire active area. Therefore, this design effectively increases the detector's detection area while maintaining low capacitance, thereby reducing series noise and enhancing the overall performance of the detector.

Meanwhile, the design and optimization process of the novel HPGeDD proposed in this paper, which has not yet been actually manufactured, is fraught with challenges and uncertainties. Due to the high cost of HPGe wafers and the large number required for preliminary experiments, blindly proceeding with actual manufacturing without adequate theoretical groundwork would result in enormous resource wastage. Therefore, conducting preliminary TCAD simulation and modeling of the detector electrical performance is particularly important.

For this purpose, the semiconductor simulation software Sentaurus TCAD was utilized in this paper to conduct electrical performance simulations of the HPGeDD. The simulation results indicate that the electric potential distribution within the detector is smooth, and the transverse electric field parallel to the electrode direction effectively facilitates the collection of induced holes, forming a hole drift channel that precisely directs toward the collecting cathode within the sensitive region. Furthermore, simulations of heavy ion incidence on the HPGeDD under operating conditions were carried out. The process of electric signal generation in the detector is observed. It has also been verified that the HPGeDD exhibits characteristics such as fast response time and short collection time. The simulation results presented in this paper not only aid in a better understanding of the structural features and working principles of the HPGeDD but also provide theoretical guidance for the future actual manufacture of HPGeDDs.

2. Detector Structure

As shown in Figure 1a, the HPGeDD is designed with a concentric ring structure. The detector substrate is made of p-type, high-purity germanium with a resistivity of 10,000 Ω·cm and a light doping concentration of $1 \times 10^{12}/\text{cm}^3$, with an overall thickness of 500 μm. As illustrated in Figure 1b, a central collecting cathode with a radius of 50 μm and a thickness of 1 μm is located at the very center of the detector. This cathode is fabricated from p-type, high-purity germanium with a heavy doping (ion implantation, etc.) concentration of $1 \times 10^{19}/\text{cm}^3$ and is primarily responsible for collecting the induced hole signals. Surrounding the central cathode are 20 concentric anode rings, arranged with an equal spacing of 30 μm. Each anode ring has a width of 80 μm and a thickness of 1 μm. The anode rings are made of n-type, high-purity germanium with a heavy doping (ion implantation, etc.) concentration of $1 \times 10^{19}/\text{cm}^3$.

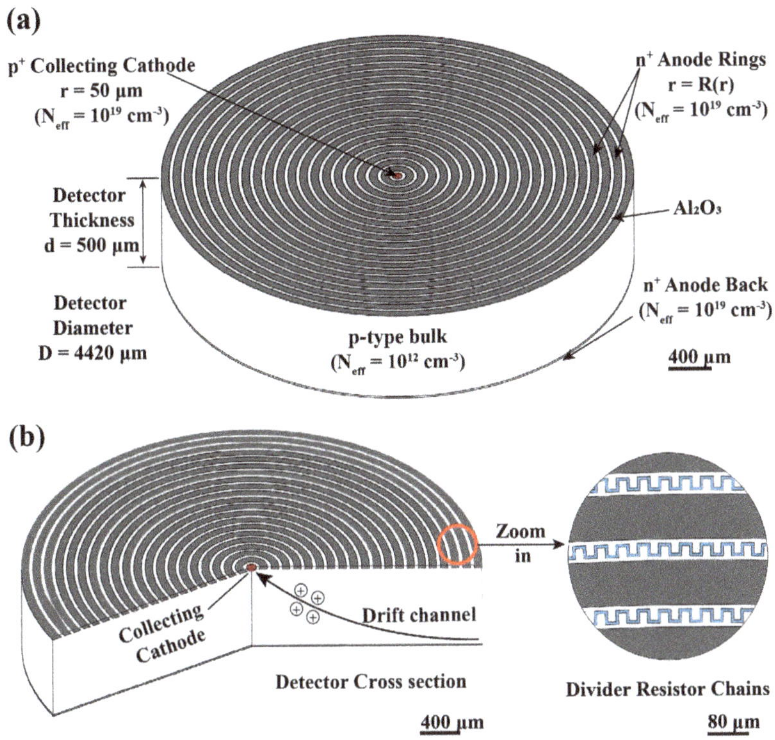

Figure 1. Schematic diagram of the HPGeDD. (**a**) Structure of the concentric anode rings HPGeDD. (**b**) Cross-sectional view of the HPGeDD and a locally enlarged view of the resistor chain.

Between adjacent anode rings, resistor chains are formed through the ion implantation of heavily doped n-type, high-purity germanium with a concentration of $1 \times 10^{19}/\text{cm}^3$ and a thickness of 1 μm is placed. Additionally, the electrode contact layer is made of aluminum, and Al_2O_3 is used as the passivation layer between the electrodes to ensure effective isolation between the electrodes.

3. Design of Voltage-Dividing Resistor Chains

As shown in Figure 2a, to achieve the autonomous voltage division function of the HPGeDD, we incorporated resistor chains between the anode rings. Figure 2b presents a schematic of a single segment in the voltage-dividing resistor chain, which is constructed by connecting multiple such segments in a series. Each resistor chain connects to two adjacent anode rings at its two ends. Since an equal increment in the voltage division is required between each anode ring, we need to add 19 resistor chains of equal length between the first anode ring and the 20th anode ring based on the actual requirements and determine the corresponding arc degrees for the resistor chains configured between the anode rings.

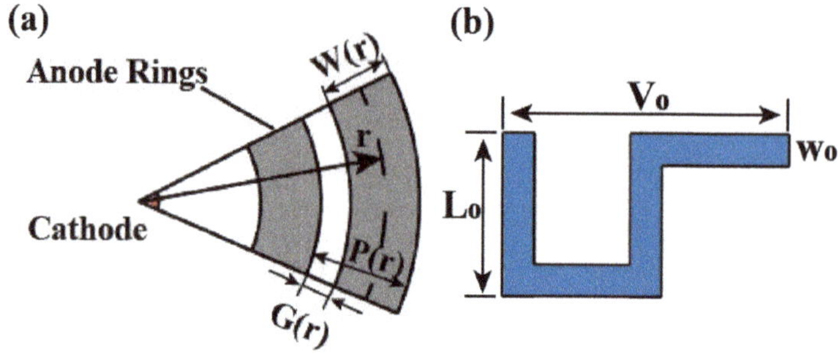

Figure 2. Design diagram of the voltage divider resistor chain for HPGeDD: (**a**) schematic diagram of the concentric anode rings for HPGeDD; (**b**) single resistor chain segment of the HPGeDD.

As shown in Figure 2, r_1 is the radial distance between the cathode center and the first anode ring, r is the radial distance between the cathode center and the r-th anode ring, r = R is the radial distance between the cathode center and the outermost 20th anode ring, $G(r)$ is the gap spacing between adjacent anode rings, $W(r)$ is the width of the anode ring, and $P(r)$ is the pitch at r between adjacent anode rings. We assumed the following in our simulation:

$$W(r) = W_0 = constant = 80\ \mu\text{m} \tag{1}$$

$$G(r) = G_0 = constant = 30\ \mu\text{m} \tag{2}$$

$$P(r) = W(r) + G(r) = 110\ \mu\text{m} \tag{3}$$

The first ring has an $r = r_1 = 120\ \mu\text{m}$, and the last ring has an $r = R = 2.21$ mm. The resistive divider chain is located between adjacent anode rings with n^+ implantation. The width of each resistor chain segment is w_0, the length is L_0, and the length of one unit is V_0. Within the interval of an anode ring with a radius of $r + \frac{1}{2}(G_0 + W_0)$, the arc length of the resistor chain corresponds to a radian measure of $\theta(r)$. $k(r)$ represents the number of segments corresponding to each resistor chain.

The total length of the resistive chain is as follows:

$$L(r) = k(r)\cdot(2L_0 + V_0) \tag{4}$$

$$\theta(r)\cdot\left[r + \frac{1}{2}(G_0 + W_0)\right] = k(r)\cdot V_0 \tag{5}$$

The resistance between the r-th ring and the $r + P_0$-th ring is as follows:

$$\Omega(r) = \frac{\rho_s\cdot L(r)}{w_0} \tag{6}$$

or

$$\Omega(r) = \frac{\rho_s\cdot k(r)\cdot(2l_0 + V_0)}{w_0} \tag{7}$$

The voltage between the r-th ring and the $r + P_0$-th ring is as follows:

$$\Delta V(r) = I\cdot\Omega(r) = \frac{I\cdot\rho_s\cdot k(r)\cdot(2l_0 + V_0)}{w_0} \tag{8}$$

The following is the total number of anode rings (N):

$$N = \frac{(R - r_1)}{P_0} + 1 \tag{9}$$

The outermost ring voltage is as follows:

$$V_{out} = 232\ \mathrm{V}(r = R) \tag{10}$$

The innermost ring voltage is as follows:

$$V_{E1} = 80\ \mathrm{V}(r = r_1) \tag{11}$$

$$\Delta V(r) = constant = \frac{(V_{out} - V_{E1})}{\left(\frac{R - r_1}{P_0}\right)} = 8\ \mathrm{V} \tag{12}$$

Appropriate values for P_0, $G_0 + W_0$, I, and $V_{out} - V_{E1}$ need to be selected to determine $k(r)$ and $\theta(r)$, as follows:

$$P_0 = G_0 + W_0 = 0.011\ \mathrm{cm} \tag{13}$$

$$I = 3 \times 10^{-7}\ \mathrm{A} \tag{14}$$

$$V_{out} - V_{E1} = 152\ \mathrm{V} \tag{15}$$

$$\rho_s = 2000\ \Omega \tag{16}$$

$$R = 2210\ \mu\mathrm{m} = 0.221\ \mathrm{cm} \tag{17}$$

Let $L_0 = 25\ \mu\mathrm{m}$, $V_0 = 50\ \mu\mathrm{m}$, and $w_0 = 5\ \mu\mathrm{m}$. Because $\theta(r_1) = 2\pi$, substituting $r = r_1$ into Equation (5) leads to the following:

$$k(r_1) = 22 \tag{18}$$

Assume the following:

$$k(r) = k(r_1) \tag{19}$$

That is, the number of segments corresponding to each resistor chain is 22. From Equation (5), we know the following:

$$\theta(r) = 2\pi \cdot \frac{k(r) \cdot V_0}{r + \frac{1}{2}(G_0 + W_0)} \tag{20}$$

The results of substituting the corresponding values into Equation (20) yields are provided in Table 1.

Table 1. The radian values correspond to resistor chains at different positions. The units for the radian values in the table are uniformly rad.

$\theta(r_1) = 6.28$	$\theta(r_2) = 3.86$	$\theta(r_3) = 2.78$	$\theta(r_4) = 2.18$
$\theta(r_5) = 1.79$	$\theta(r_6) = 1.52$	$\theta(r_7) = 1.32$	$\theta(r_8) = 1.16$
$\theta(r_9) = 1.04$	$\theta(r_{10}) = 0.94$	$\theta(r_{11}) = 0.86$	$\theta(r_{12}) = 0.79$
$\theta(r_{13}) = 0.74$	$\theta(r_{14}) = 0.69$	$\theta(r_{15}) = 0.64$	$\theta(r_{16}) = 0.60$
$\theta(r_{17}) = 0.57$	$\theta(r_{18}) = 0.54$	$\theta(r_{19}) = 0.51$	

By adding the corresponding arc length of the resistor chain in the appropriate positions based on the aforementioned radian values, $\theta(r)$, a voltage division effect can be achieved.

4. Simulation Results of Detector Electrical Performance

This study employs the semiconductor simulation software Sentaurus TCAD (P-2019.03) to conduct an in-depth investigation into the electrical performance of the HPGeDD. We utilize carrier transport equations, the SRH (Shockley–Read–Hall) recombination model, and the density of states model to analyze this device. For material parameters, we directly adopt those built into the Sentaurus TCAD software. In the mobility model, we consider carrier scattering by ionized impurities and incorporate the carrier mobility degradation model (DopingDep), the carrier velocity saturation model in high electric fields (HighFieldSat), and the mobility degradation model due to surface roughness scattering (Enormal). Regarding the density of states model, taking into account the doping effects during the simulation process, we employ the effective intrinsic density model (OldSlotboom). In the recombination model, the corresponding carrier generation and recombination processes are implemented through the carrier continuity equation, with the SRH recombination mechanism being selected. Additionally, an interface charge density of -4×10^{11} q/cm^2 is set between Al_2O_3 and Ge.

To further enhance the accuracy of our simulations, we carefully defined the mesh size for the entire device. Specifically, the maximum mesh setting for the x/y/z coordinates is 8 μm, while the minimum mesh setting is 3 μm. Moreover, we refined the mesh at material interfaces and locations where the doping concentration changes. This refinement starts with an initial mesh size of 2 μm and increments it by a factor of 1.5 to ensure a more precise simulation of these critical areas.

As shown in Figure 3a, a three-dimensional (3D) image of the HPGeDD is displayed. Since the model adopted in this study is a concentric ring detector with a highly symmetrical structure, we selected a representative tangential position at half of the detector's cross-section for investigation. In the simulation experiment, a voltage of 0 V is applied to the central cathode, 80 V to the first anode ring, 232 V to the last anode ring, and 152.5 V to the bottom anode plate.

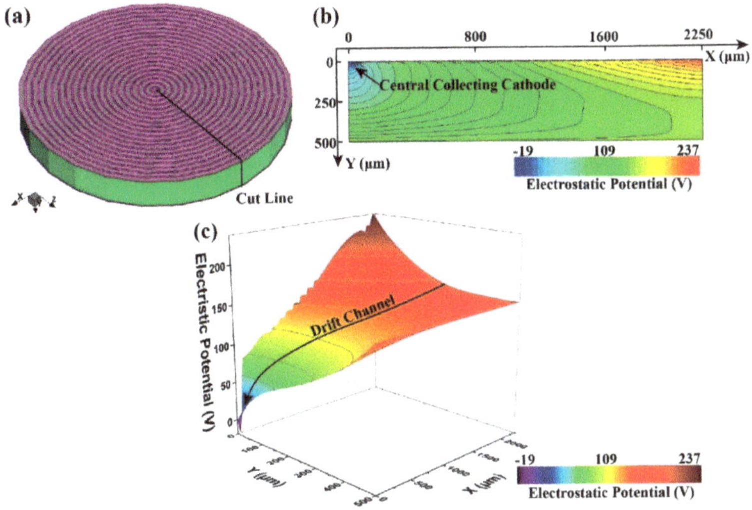

Figure 3. Electrical simulation images of the HPGeDD: (**a**) 3D simulation image of the HPGeDD, with a two-dimensional cross-section taken at a tangential position; (**b**) potential distribution map of the two-dimensional cross-section; (**c**) 3D potential distribution map of the HPGeDD. The drift channel is marked with a black line with an arrow.

Figure 3b presents the electric potential distribution of the HPGeDD. In the upper half of the detector, the potential is lowest in the collection cathode area located at the center on the left side, and it gradually increases from left to right. In the lower half, the potential of the bottom anode plate remains constant. By further observing Figure 3c, which shows the 3D potential distribution across a cross-section of the detector, we can see that a low-potential valley forming a carrier drift channel within the sensitive area of the detector. This channel extends smoothly from the edge of the detector to the central collection cathode (as indicated by the black arrow in the figure), providing favorable conditions for holes within the substrate to move toward and be absorbed by the central cathode.

When induced holes are generated inside the detector, they preferentially drift into this low-potential hole drift channel. Subsequently, the holes move gradually toward the central collection cathode along the direction of decreasing potential within the drift channel and are ultimately collected. Additionally, the potential changes continuously and smoothly around the anode rings, with no abrupt potential variations. This allows for the holes excited by incident particles to drift accurately into the drift channel and be collected. These results indicate that the design presented in this paper has achieved the expected goals.

To further explore the depletion process within the HPGeDD, as well as the formation mechanism of the electric field and the hole drift channel, we recorded the distributions of electric field and hole concentration of the detector under different voltages. Figure 4a presents the electric field distribution within the detector under various biasing conditions. As the voltage gradually increases, the electric field strength in the upper half of the detector enhances, gradually pushing the low-field regions toward the vicinity of the drift channel. Simultaneously, the electric field in the lower half of the detector also intensifies with increasing voltage, similarly squeezing the low-field regions toward the central drift channel. When the detector reaches full depletion, the smooth electric field within its sensitive area effectively promotes the separation and drift of electron–hole pairs generated by radiation ionization, thereby generating an electrical signal. The induced holes will move toward the drift channel under the action of the electric field. Inside the channel, there is a gradually increasing electric field pointing toward the collection cathode, and the induced holes continue to drift to the central collection cathode under the influence of this electric field.

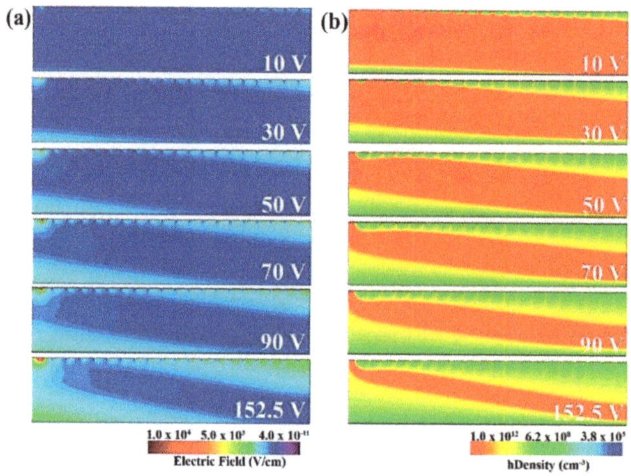

Figure 4. Distributions of the internal electric field and hole density in the HPGeDD as a function of voltage, with bottom anode plate biases from 10 V to 152.5 V: (**a**) electric field; (**b**) hole density.

The distributions of the detector's hole concentrations under different biasing voltage conditions are presented in Figure 4b. As the biasing voltage gradually increases, the hole depletion zones extend from both surfaces toward the middle of the detector, gradually forming a hole drift channel, most visible at a bias of 152.5 V. It is noteworthy that the path for holes in the carrier drift channel to transport to the central collection cathode is the shortest. This short drift path effectively reduces the diffusion and recombination losses of holes during the drift process, thereby improving the detector's collection efficiency and energy resolution.

As shown in Figure 5, the curve of the leakage current versus voltage for the HPGeDD is presented. The leakage current of the detector primarily originates from impurities and lattice defects in the material. These impurities and defects give rise to the generation–recombination current that is the main contributor to the detector leakage current. Before the detector reaches full depletion, as the reverse bias voltage gradually increases, the depletion layer expands. Since the detector leakage current is proportional to the depletion depth, the leakage current increases linearly with the reverse bias voltage. When the reverse bias voltage reaches a certain threshold, the depletion layer extends to the entire detector thickness, at which point the detector reaches full depletion, the leakage current also reaches its maximum value (about 8×10^{-7} A). The magnitude of the leakage current directly affects the sensitivity and resolution of the detector. A larger leakage current increases the noise level of the detector, thereby reducing its signal-to-noise ratio. Therefore, controlling the leakage current is a crucial factor in improving detector performance.

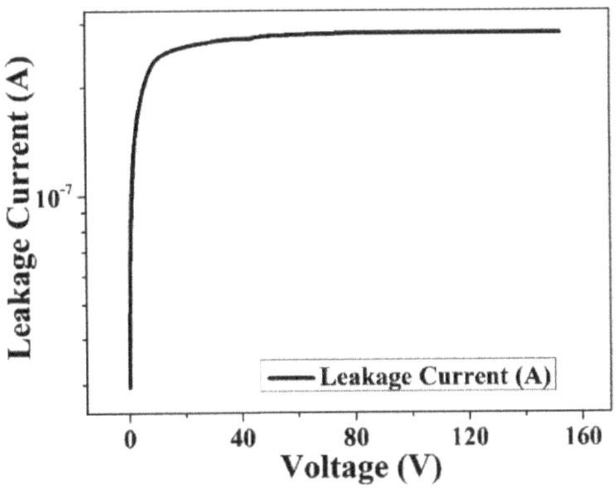

Figure 5. Leakage current versus voltage curve for an HPGeDD.

As shown in Figure 6, the graph illustrates the variation in capacitance with voltage for an HPGeDD. As the applied voltage gradually increases, the depletion region of the detector expands accordingly. Since the capacitance of the detector is inversely proportional to the thickness of the depletion depth, the capacitance of the detector gradually decreases as the depletion region expands. When the applied voltage reaches a full depletion voltage, the depletion depth is the detector thickness, and the capacitance no longer undergoes significant changes and stabilizes at 0.2 fF. The fully depleted state is the optimal state for detector performance because, in this state, all carriers within the entire volume can be efficiently collected, enabling highly effective signal response. The capacitance of the high-purity germanium drift detector proposed in this paper stabilizes at 0.2 fF in the fully depleted state, which is significantly smaller than the reported capacitance of 1 pF for point-contact electrode, high-purity germanium detectors [23], demonstrating a notable advantage of our detectors in capacitance characteristics.

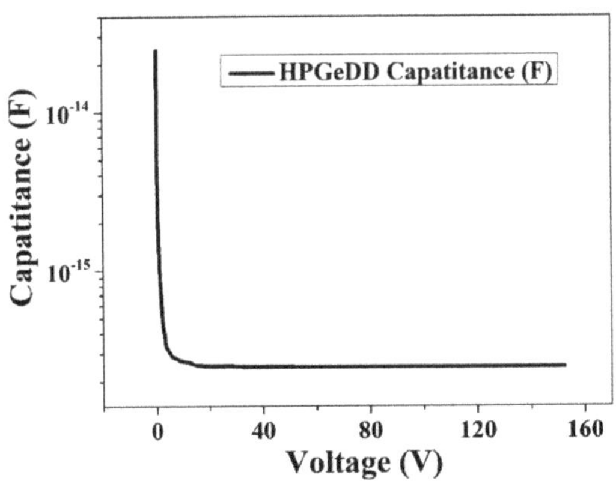

Figure 6. Capacitance versus voltage curve for the HPGeDD.

5. Heavy Ion Incident Simulation

To investigate the carrier collection process of the HPGeDD under operating conditions, we conducted a simulation experiment involving heavy ion incidence on the detector. When the detector reached the fully depleted state, we selected a position at 750 μm along the x-axis as the incidence location, and a beam of heavy ions was vertically injected from the bottom of the detector. The initial time of incidence is at t = 5 ns, with an incidence depth of 200 μm. The radius of the interaction zone is set at 2 μm, and the linear energy transfer rate (LET) is set to 1×10^{-5} pC/μm.

As shown in Figure 7, the image depicts the variation in hole density over time within the HPGeDD after the heavy ion incidence. Through a series of continuous images from 0 ns to 300 ns, we can clearly observe that when the heavy ions strike the detector, it deposits energy at the corresponding location. This energy deposition process subsequently excites the generation of electron–hole pairs within the sensitive area of the detector. Under the influence of the electric field, the holes move toward the hole drift channel and then toward the central collecting cathode, where they are eventually collected to form an electric signal. The continuous change in hole density over time in the simulation image showcases the entire process from the generation to the collection of induced holes. At 300 ns, the detector completes the charge collection and returns to its initial state. Throughout the entire process, from the moment the heavy ion incidence to the final completion of charge collection, the detector continuously generates a corresponding induced current signal.

As shown in Figure 8, the simulation of the induced currents by heavy ion incidence on the HPGeDD at different positions within the detector is presented, observing the current variation over time after the heavy ions enters the detector. For a single curve, at t = 5 ns, the moment of heavy ion incidence, the current value begins to appear and gradually increases. As the carriers are gradually transported to the central collecting cathode, the current value continues to rise until it reaches a peak. Subsequently, as a large number of carriers are collected by the central collecting cathode, the current gradually decreases to the lowest induced current level and eventually returns to the depleted state. By comparing different curves, it can be observed that positions closer to the central collecting cathode exhibit larger current peak signals, reach the peak earlier, and have the shortest signal collection time. This is because the distance for induced carriers to drift to the collecting cathode is shorter at positions closer to the cathode, resulting in a shorter collection time. Conversely, at positions farther away, the drift distance for induced carriers is longer, leading to a correspondingly longer collection time, a lower peak value, and a delayed time to reach the peak.

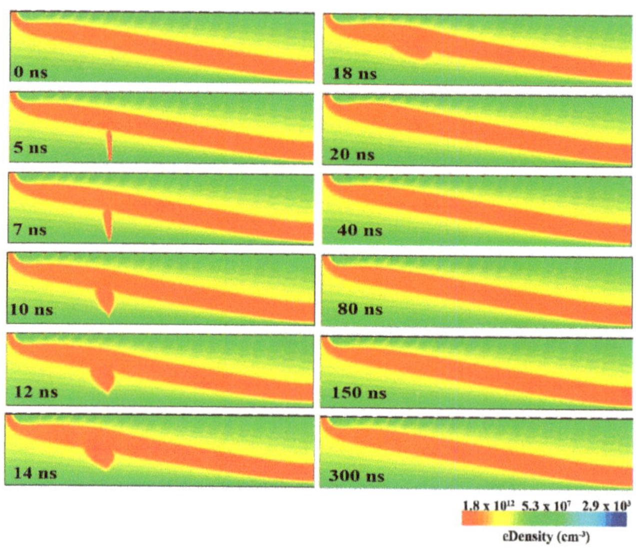

Figure 7. Simulation of heavy ion incidence on the HPGeDD. The incident time is 5 ns, and the incident position is x = 750 μm.

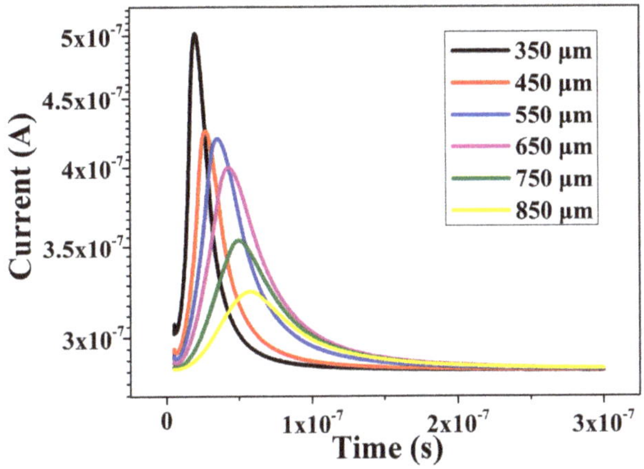

Figure 8. Curves showing the relationship between the current and time at the central cathode of the HPGeDD under heavy ion incidence at different positions.

The simulation results above indicate that when using an HPGeDD, we should be mindful of the variations in signal response at different positions, with locations nearer to the central collecting cathode exhibiting higher signal amplitude and faster response times. This characteristic necessitates that we thoroughly consider the impact of positional effects on measurement outcomes when designing experiments or analyzing data. Furthermore, for measurement tasks requiring high-precision time resolution or strong signal peaks, it is advisable to prioritize detecting in the region close to the central collecting cathode to optimize detection efficiency and data quality.

6. Conclusions

Traditional HPGe detectors that aim to detect a broader range of radiation are often manufactured with centimeter-scale dimensions to achieve ultra-large sizes. However, the large size of HPGe detectors inevitably introduces the side effect of increased capacitance. Detector capacitance is a key factor affecting the noise level of the detector, and maintaining a low noise level is crucial for a stable detector performance. Reducing the overall size of the detector to decrease its capacitance would introduce a new issue with the reduction of the active area of the detector. On the other hand, the large volume of traditional HPGe detectors is not conducive to their application in aerospace and other fields where portability and small size are required.

To address this issue, this paper proposes an HPGeDD with a concentric ring structure based on the existing structure of drift detectors. The capacitance of this detector is only related to the area of the central collecting cathode, rather than the active area of the detector. Therefore, it is possible to significantly increase the overall area of the detector while keeping its capacitance low, thereby expanding the detection area of the detector. Electrical performance simulations of this detector structure were conducted using the semiconductor simulation software Sentaurus TCAD. The simulation results show that the potential distribution inside the detector is smooth, with no abrupt changes, and there is a lateral drift electric field parallel to the detector surfaces. The carrier drift channel pointing toward the central collecting cathode is clearly visible. Our detector has shown a low leakage current of 8×10^{-7} A and an extremely low capacitance of 0.2 fF. Furthermore, heavy ion incidence simulations were performed on the HPGeDD. The simulation results indicate that holes move along the carrier transport channel in the middle of the detector toward the central collecting cathode and are ultimately collected to generate an electrical signal. From the comparison of current signal curves, it can be observed that the closer the incidence position is to the center of the cathode, the shorter the collection time and the stronger the electrical signal.

To realize the HPGeDD proposed in this paper, key technologies that may be employed include atomic layer deposition (ALD) technology for ensuring uniform and pure deposition of Al_2O_3 on the wafer surface; ion implantation technology for precise control of doping concentrations; and microfabrication techniques such as photolithography and etching for fabricating the detector's intricate structures. Additionally, packaging technology is required to protect the detector from external environmental interference, and a rigorous testing process is necessary to ensure stable and reliable detector performance, meeting the requirements for high-sensitivity and high-resolution detection.

Furthermore, we have outlined our next research steps, which involve conducting a systematic study on the radiation damage effects of HPGeDDs. This research aims to delve deeper into the mechanisms of radiation damage, exploring how external factors such as radiation dose and exposure time influence the performance of semiconductor detectors. Understanding these effects is crucial for the further development and optimization of HPGeDDs, and it represents an important extension of our current research efforts. Therefore, we plan to dedicate our upcoming work to this systematic study.

The research content of this paper is crucial for the subsequent development of HPGeDDs, as it can effectively optimize the performance of HPGe detectors and provide possibilities for their application in a wider range of fields.

Author Contributions: Conceptualization, M.W. and Z.L.; data curation, M.W.; formal analysis, M.W. and Z.L.; funding acquisition, Z.L. and Y.X.; investigation, M.W. and B.X.; methodology, M.W.; project administration, Z.L. and Y.X.; resources, Z.L. and Y.X.; software—Sentaurus TCAD (P-2019.03), M.W. and B.X.; supervision, Z.L. and Y.X.; validation, Z.L. and Y.X.; visualization, M.W. and Z.L.; writing—original draft, M.W.; writing—review and editing, M.W., Z.L. and Y.X. All authors have read and agreed to the published version of the manuscript.

Funding: The authors would like to express their gratitude for the financial support received from the Key Project of National Natural Science Foundation of China (11835008) and the National Key R&D Program of China (2023YFF0719600), National Natural Science Foundation of China (Grant Nos. 92164108 and 11835008), Key Project of Scientific Research Fund of Hunan Provincial Education Department (Grant No. 23A0150), and Project of Hunan Provincial NSFC (Grant Nos. 2023JJ50009 and 2023JJ30599). Additionally, this work was supported by Postgraduate Scientific Research Innovation Project of Hunan Province (CX20220553).

Data Availability Statement: The data presented in this study are available from the corresponding author upon reasonable request.

Conflicts of Interest: The authors declare no conflicts of interest.

References

1. Hubbard, G.S.; Haller, E.E.; Hansen, W.L. Zone refining high-purity germanium. *IEEE Trans. Nucl. Sci.* **1978**, *25*, 362–370. [CrossRef]
2. Pascovici, G.; Pullia, A.; Zocca, F.; Bruyneel, B.; Bazzacco, D. Low noise, dual gain preamplifier with built in spectroscopic pulser for highly segmented high-purity germanium detectors. *Wseas Trans. Circuits Syst.* **2008**, *7*, 470–481.
3. Amman, M.; Looker, Q.; Vetter, K. Inter-electrode charge collection in high-purity germanium detectors with amorphous semiconductor contacts. *Nucl. Instrum. Methods Phys. Res. Sect. A Accel. Spectrometers Detect. Assoc. Equip.* **2015**, *781*, 20–25.
4. Qian, Y.; Cheng, J.-P.; Li, Y.-J.; Li, J.; Wang, Z.-J. Detection of wimps using low threshold hpge detector. *High Energy Phys. Nucl. Phys.* **2004**, *28*, 877–880.
5. He, D.; Li, Y.; Yue, Q.; Li, Y.; Li, J.; Cheng, J.; Kwak, J.W.; Kim, S.K.; Li, X.; Wang, H.T. Experimental study on the csi (tl) crystal anti-compton detector for dark matter search. *High Energy Phys. Nucl. Phys.* **2006**, *30*, 548–553.

6. Kang, K.J.; Cheng, J.P.; Li, J.; Li, Y.J.; Yue, Q.; Bai, Y.; Bi, Y. Introduction to the cdex experiment. *Front. Phys.* **2013**, *8*, 412–437. [CrossRef]
7. Antonopoulou, D.; Haskell, B. Pulsar glitches: Observations and physical interpretation. *Rep. Prog. Phys.* **2022**, *85*, 126901. [CrossRef]
8. Liu, T.; Lou, X.; Ren, J. Pulsar polarization arrays. *Phys. Rev. Lett.* **2023**, *130*, 121401. [CrossRef]
9. Pan, Z.; Lu, J.G.; Jiang, P.; Han, J.L.; Chen, H.L.; Han, Z.W.; Liu, K.; Qian, L.; Xu, R.X.; Zhang, B. A binary pulsar in a 53-minute orbit. *Nature* **2023**, *620*, 961–964. [CrossRef]
10. Philippov, A.; Timokhin, A.; Spitkovsky, A. Origin of pulsar radio emission. *Phys. Rev. Lett.* **2020**, *124*, 245101. [CrossRef]
11. Li, D.; Bilous, A.; Ransom, S.; Main, R.; Yang, Y.P. A highly magnetized environment in a pulsar binary system. *Nature* **2023**, *618*, 484–488. [CrossRef] [PubMed]
12. Zhang, L.; Fang, H.; Zhao, S.L. Enhanced x-ray pulsar navigation based on ranging information of a satellite. *Aircr. Eng. Aerosp. Technol.* **2022**, *94*, 1872–1882. [CrossRef]
13. Kabakchiev, H.; Behar, V.; Kabakchieva, D.; Kisimov, V.; Stefanova, K. Investigation of accuracy of toa and snr of radio pulsar signals for vehicles navigation. *Sensors* **2023**, *23*, 7010. [CrossRef]
14. Song, S.; Xu, L.; Zhang, H.; Bai, Y. Novel X-ray communication based xnav augmentation method using X-ray detectors. *Sensors* **2015**, *15*, 22325–22342. [CrossRef]
15. Xue, M.; Shi, Y.; Guo, Y.; Huang, N.; Peng, D.; Luo, J.A.; Shentu, H.; Chen, Z. X-ray pulsar-based navigation considering spacecraft orbital motion and systematic biases. *Sensors* **2019**, *19*, 1877. [CrossRef]
16. Chen, Q.; Zhao, Y.; Yan, L. X-ray pulsar signal denoising based on variational mode decomposition. *Entropy* **2021**, *23*, 1181. [CrossRef]
17. Ely, T.; Bhaskaran, S.; Bradley, N.; Lazio, T.J.W.; Martin-Mur, T. Comparison of deep space navigation using optical imaging, pulsar time-of-arrival tracking, and/or radiometric tracking. *J. Astronaut. Sci.* **2022**, *69*, 385–472. [CrossRef]
18. Wang, Y.; Zheng, W.; Zhang, S.; Minyu, G.E.; Liansheng, L.I.; Jiang, K.; Chen, X.; Zhang, X.; Zheng, S.; Fangjun, L.U. Review of x-ray pulsar spacecraft autonomous navigation. *Chin. J. Aeronaut.* **2023**, *36*, 44–63. [CrossRef]
19. Yang, M.; Li, Y.; Zeng, Z.; Tian, Y.; Xue, T.; Zeng, M.; Tang, W.; Chang, J. A prototype segmented planar high purity germanium detector using wraparound lithium diffusion electrode and amorphous germanium blocking contact. *Radiat. Detect. Technol. Methods* **2022**, *6*, 433–438. [CrossRef]
20. Gladen, R.W.; Harvey, T.J.; Chirayath, S.S.; Fairchild, A.J.; Koymen, A.R.; Weiss, A.H.; Chirayath, V.A. Implementation of a machine learning technique for estimating gamma direction using a coaxial high purity germanium detector. *Nucl. Instrum. Methods Phys. Res. Sect. A Accel. Spectrometers Detect. Assoc. Equip.* **2022**, *1039*, 167067. [CrossRef]
21. Lin, S.T.; Li, H.B.; Li, X.; Lin, S.K.; Wong, H.T.; Deniz, M.; Fang, B.B.; He, D.; Li, J.; Lin, C.W.; et al. New limits on spin-independent and spin-dependent couplings of low-mass wimp dark matter with a germanium detector at a threshold of 220 ev. *Phys. Rev. D* **2009**, *79*, 061101. [CrossRef]
22. Li, Y.Y.; Xiong, B.; Li, Z. 3D design and electric simulation of a silicon drift detector using a spiral biasing adapter. *Nucl. Instrum. Methods Phys. Res.* **2016**, *831*, 29–33.
23. Eberth, J.; Simpson, J. From ge(li) detectors to gamma-ray tracking arrays–50 years of gamma spectroscopy with germanium detectors. *Prog. Part. Nucl. Phys.* **2008**, *60*, 283–337. [CrossRef]

micromachines

MDPI

Article

Investigation of Modal Characteristics of Silicon Nitride Ridge Waveguides for Enhanced Refractive Index Sensing

Muhammad A. Butt [1,*], Lukasz Kozlowski [1], Mateusz Słowikowski [2], Marcin Juchniewicz [2], Dagmara Drecka [2], Maciej Filipiak [2], Michał Golas [2], Bartłomiej Stonio [2], Michal Dudek [3] and Ryszard Piramidowicz [1]

1 Institute of Microelectronics and Optoelectronics, Warsaw University of Technology, Koszykowa 75, 00-662 Warsaw, Poland
2 The Centre for Advanced Materials and Technologies CEZAMAT, Warsaw University of Technology, Poleczki 19, 02-822 Warsaw, Poland
3 Institute of Applied Physics, Military University of Technology, Gen. Sylwestra Kaliskiego 2, 00-908 Warsaw, Poland
* Correspondence: ali.butt@pw.edu.pl

Abstract: This paper investigates the wavelength-dependent sensitivity of a ridge waveguide based on a silicon nitride (Si_3N_4) platform, combining numerical analysis and experimental validation. In the first part, the modal characteristics of a Si_3N_4 ridge waveguide are analyzed in detail, focusing on the effective refractive index (n_{eff}), evanescent field ratio (EFR), and propagation losses (α_{prop}). These parameters are critical for understanding the interplay of guided light with the surrounding medium and optimizing waveguide design for sensing applications. In the second part, the wavelength-dependent sensitivity of a racetrack ring resonator (RTRR) based on the Si_3N_4 waveguide is experimentally demonstrated. The results demonstrate a clear increase in the sensitivity of the RTRR, rising from 116.3 nm/RIU to 143.3 nm/RIU as the wavelength shifts from 1520 nm to 1600 nm. This trend provides valuable insights into the device's enhanced performance at longer wavelengths, underscoring its potential for applications requiring high sensitivity in this spectral range.

Keywords: silicon nitride; ridge waveguide; racetrack ring resonator; sensitivity enhancement; effective refractive index

1. Introduction

Silicon nitride (Si_3N_4) photonics is an essential platform in modern integrated photonics that is known for its versatility and exceptional optical properties [1,2]. Si_3N_4 offers a wide optical transparency range, extending from the visible spectrum to the MIR (400 nm to 6.7 μm) [3], making it suitable for diverse applications such as telecommunications, biosensing, quantum computing, and LiDAR [4–7]. One of its primary advantages is its ultra-low optical loss, particularly in the near-infrared region, where losses can be less than 1 dB/cm [8]. The material's moderate refractive index (~2.0) enables efficient mode confinement while maintaining robustness against fabrication imperfections unlike higher contrast materials such as silicon [9]. Its compatibility with CMOS fabrication processes ensures scalability and cost-efficiency, facilitating seamless integration with silicon photonics [10,11]. Si_3N_4 also exhibits high thermal stability and a low thermo-optic coefficient, making it ideal for temperature-sensitive applications [12]. These features collectively position Si_3N_4 photonics as a key enabler for next-generation photonic integrated circuits (PICs), addressing the needs of both research and industry [13,14].

Optical waveguides are physical structures that guide light waves along a specific path, and they are often used in telecommunications, sensors, and photonics [15]. They confine and direct light through reflection and refraction, typically in the form of optical fibers or planar waveguides [16]. These waveguides play a crucial role in modern technology, supporting high-speed data transmission and advanced optical systems. The sensitivity of a waveguide is intrinsically tied to its dimensions because these determine how the electromagnetic field of the guided mode interacts with the surrounding environment [17,18]. In a waveguide, the dimensions influence the mode profile, effective index (n_{eff}), and evanescent field strength [19,20]. Waveguide sensitivity is significantly influenced by its cut-off wavelength, which is determined by the physical geometry and size of the waveguide [21,22]. This sensitivity to cut-off wavelength makes waveguides ideal for selective wavelength transmission but also introduces design constraints, as the waveguide must be precisely engineered to accommodate the desired wavelength range [23].

To design highly sensitive photonic sensors, it is essential to maximize the overlap of the guided mode's electric field with the surrounding ambient medium. This enhanced overlap upsurges the interaction between the optical mode and the external environment, significantly improving the sensor's ability to detect changes in the medium, for instance, variations in refractive index (RI), chemical concentrations, or other perturbations [24,25]. However, achieving this greater interaction often comes at the cost of increased waveguide losses. These losses can result from several factors, including radiation losses due to energy leakage into unguided modes, scattering losses caused by imperfections at the waveguide interface, and material absorption within the ambient medium [26]. In this paper, we comprehensively analyze the modal characteristics of the Si_3N_4 waveguide, focusing on the evanescent field ratio (EFR), propagation loss (α_{prop}), and sensitivity across a broad wavelength spectrum. This investigation offers valuable insights into the relationship between sensitivity and wavelength. Furthermore, we experimentally demonstrate this phenomenon using a Si_3N_4-based RTRR, measuring its sensitivity across six distinct wavelength regions.

2. Waveguide Design and Numerical Analysis

The ridge waveguide consists of a Si_3N_4core deposited on a SiO_2 on Si substrate as shown in Figure 1. The core dimensions are characterized by their height (H_{core}) and width (W_{core}). The geometric parameters utilized for the waveguide analysis are detailed in Table 1. The refractive indices of Si_3N_4 and SiO_2 were sourced from the well-established works of Luke et al. [27] and Kischkat et al. [28], respectively, ensuring accurate and reliable material parameters for the analysis. COMSOL Multiphysics 5.5 is a powerful simulation software widely used for analyzing optical waveguide-based photonic devices. It provides a versatile platform to model the electromagnetic behavior of photonic structures by solving Maxwell's equations with high accuracy. By employing the Wave Optics Module, we analyzed critical parameters such as mode profiles, effective refractive indices (n_{eff}), propagation losses (α_{prop}), and wavelength-dependent behavior [29].

Table 1. Description of the geometric parameters used in the analysis of a ridge waveguide based on Si_3N_4 material platform.

Variable	Description	Range
W_{core}	Width of a waveguide core	800 nm to 1200 nm
H_{core}	Height of a waveguide core	340 nm to 400 nm

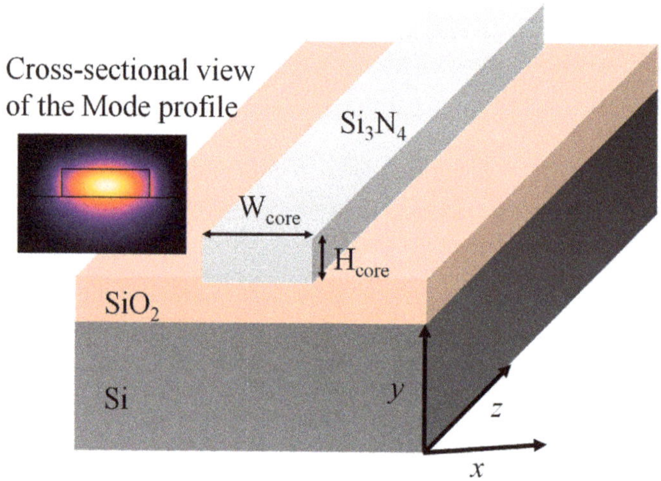

Figure 1. Graphic of a Si_3N_4 ridge waveguide. The inset shows the norm. E-field distribution at 1550 nm.

At first, we employed boundary mode analysis in COMSOL Multiphysics 5.5 software to determine the n_{eff} of a waveguide, which can have both real (Re) and imaginary (Im) components. This analysis involves solving the electromagnetic (EM) field equations at the boundaries of the waveguide cross-section to identify the supported modes and their corresponding effective refractive indices. By defining the waveguide's geometry, material properties, and operating wavelength, COMSOL calculates the distribution of the electric and magnetic fields in the waveguide. The $Re(n_{eff})$ represents the phase velocity of the propagating mode relative to the speed of light in a vacuum and determines the wave's propagation characteristics within the waveguide. Figure 2a reveals that as the core size increases, $Re(n_{eff})$ rises, indicating stronger light confinement within the waveguide core.

The $Im(n_{eff})$, on the other hand, quantifies the attenuation of the mode as it propagates, indicating losses due to absorption, scattering, or leakage. The α_{prop} measured in dB/cm is linked to $Im(n_{eff})$ through the following Equation (1). The accurate determination of both components is crucial for designing efficient photonic devices, as the real part ensures proper confinement and guidance of light, while the imaginary part provides insights into energy losses.

$$\alpha\left(\frac{dB}{cm}\right) = \left(\frac{4 \times \pi \times Im(n_{eff})}{\lambda}\right) \times 4.343; \quad (1)$$

Here, λ represents the wavelength of the propagating light. This equation demonstrates that a higher $Im(n_{eff})$ leads to increased attenuation per unit length. Figure 2b shows that as the waveguide dimensions increase, α_{prop} decreases, leading to a reduction in mode leakage to the substrate. This is further corroborated by the waveguide's figure of merit (FOM), which indicates enhanced performance and improved confinement as the dimensions grow.

The ratio of mode confinement in the core to power leakage into the substrate can be considered an FOM for evaluating the performance of an optical waveguide, especially in contexts where guiding efficiency and minimizing losses are critical (Figure 2c). It is determined by performing a surface integral of the electric field intensity squared over the core region and dividing it by the surface integral of the electric field intensity squared over the substrate region. In this case, the FOM would quantify how effectively the waveguide confines the optical mode within the core relative to the unwanted leakage into the substrate. Mathematically, it could be expressed in the form of Equation (2):

$$\mathrm{FOM} = \frac{\mathrm{Modeconfinementincore}}{\mathrm{Modeleakagetosubstrate}} = \frac{\int_{\mathrm{core}} \mathrm{E}^2 \mathrm{dA}}{\int_{\mathrm{substrate}} \mathrm{E}^2 \mathrm{dA}}; \quad (2)$$

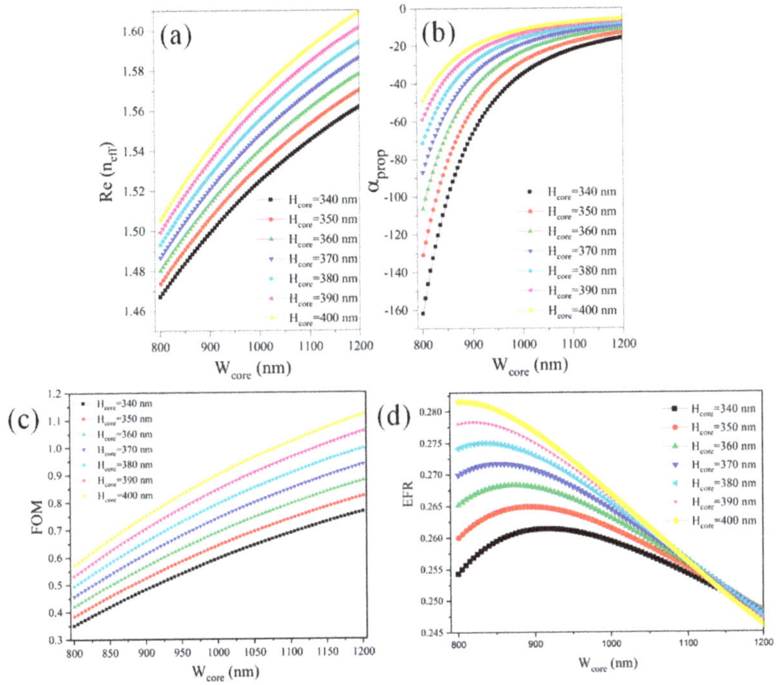

Figure 2. (**a**) Re(n_{eff}) versus waveguide dimensions, (**b**) α_{prop} versus waveguide dimensions, (**c**) FOM versus waveguide dimensions, and (**d**) EFR versus waveguide dimensions.

The FOM of the Si_3N_4 ridge waveguide improves as the waveguide dimensions increase. This enhancement is primarily attributed to a reduction in mode power leakage to the substrate and a decrease in the evanescent field. These factors, in turn, lower the Evanescent Field Ratio (EFR) of the waveguide, leading to a decrease in the device's sensitivity. The EFR quantifies the relative strength of the evanescent field compared to the total field within an optical waveguide [30]. For a ridge waveguide, the EFR is defined as the ratio of the power or intensity of the evanescent field in the upper cladding to the total field intensity, which includes the contributions from the substrate, upper cladding, and core. The EFR is expressed in the form of Equation (3).

$$\mathrm{EFR} = \frac{\iint_{\mathrm{uppercladding}} |\mathrm{E}(\mathrm{x},\mathrm{y})|^2 \mathrm{dxdy}}{\iint_{\mathrm{total}} |\mathrm{E}(\mathrm{x},\mathrm{y})|^2 \mathrm{dxdy}}; \quad (3)$$

Figure 3(a-1)–(c-3) illustrates the normalized E-field distribution in the Si_3N_4 ridge waveguide for core dimensions ranging from W_{core} = 800 nm to 1200 nm and H_{core} = 360 nm to 400 nm. As the waveguide dimensions increase, Re(n_{eff}) rises due to better mode confinement, while Im(n_{eff}) decreases. This reduces power leakage to the substrate and minimizes the evanescent field surrounding the waveguide core. For W_{core} =

800 nm and H_{core} = 400 nm, the EFR is 0.282, which decreases to 0.246 when the dimensions are expanded to W_{core} = 1200 nm and H_{core} = 400 nm. These results highlight the importance of optimizing waveguide geometry to achieve superior light confinement and minimal propagation losses.

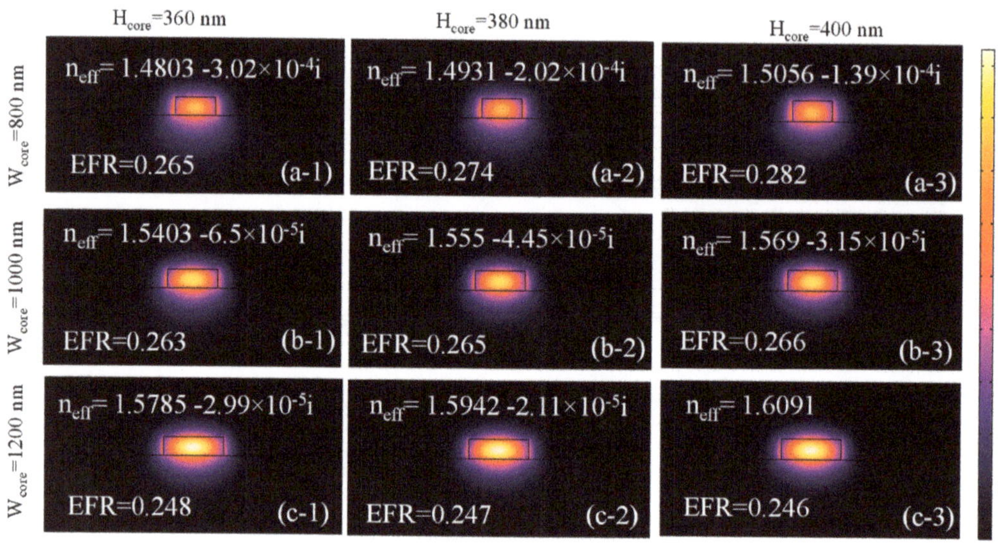

Figure 3. Norm. E-field distribution in the Si_3N_4 ridge waveguide at an operational wavelength = 1550 nm for (**a-1,b-1,c-1**) H_{core} = 360 nm and W_{core} = 800 nm, (**a-2,b-2,c-2**) H_{core} = 380 nm and W_{core} = 1000 nm, and (**a-3,b-3,c-3**) H_{core} = 400 nm and W_{core} = 1200 nm.

For sensors operating based on the wavelength interrogation method, it is crucial to evaluate both the EFR and the sensitivity of the waveguide (S_W) across a broad wavelength spectrum. For this analysis, the waveguide dimensions are fixed at W_{core} = 1000 nm and H_{core} = 400 nm. This selection is based on the specific design parameters of the RTRR, which is fabricated and analyzed in detail in the subsequent section. These dimensions were selected to align with the experimental setup, allowing for a clear and accurate comparison between theoretical predictions and experimental results. The EFR of the waveguide over a wavelength range from 1520 nm to 1600 nm was analyzed, as depicted in Figure 4a. The EFR of the waveguide increases linearly with the operational wavelength. The RI measurement range of 1.414 to 1.426 was chosen for presentation in numerical and experimental studies because this range corresponds to the refractive indices of many bioanalytes, making it particularly relevant for applications in biological and chemical sensing. Moreover, a higher ambient RI results in a more pronounced increase in the EFR, highlighting the influence of the surrounding medium on the waveguide's performance. This comprehensive assessment provides valuable insights into the waveguide's performance across the spectrum.

The sensitivity of an optical waveguide quantifies the ability of the waveguide to detect changes in the RI of the surrounding medium as presented in Equation (4).

$$S_W = \frac{\Delta n_{eff}}{\Delta n}; \quad (4)$$

Here, Δn_{eff} represents the variation in the effective refractive index of the guided mode caused by the variation in the surrounding medium's refractive index (Δn). A higher sensitivity indicates that even a small Δn produce a significant shift in n_{eff}, enhancing the detection capabilities of the waveguide. Sensitivity counts on several factors, including the waveguide geometry, the refractive index contrast between the core and the cladding, and the degree of mode confinement. From Figure 4b, it is observed that S_W of the waveguide for $\Delta n = 0.012$ is approximately 0.21 at an operational wavelength of 1520 nm. This sensitivity increases progressively, reaching around 0.235 as the wavelength approaches 1600 nm. This trend indicates an enhancement in the interplay between the guided mode and the adjacent medium at longer wavelengths, which is a critical factor for optimizing the waveguide's performance in sensing applications.

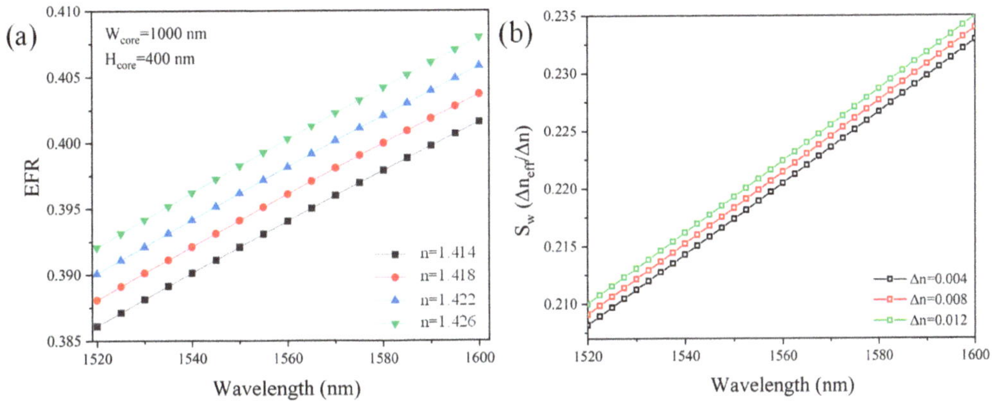

Figure 4. (**a**) EFR versus wavelength, (**b**) S_W versus wavelength. Note: In this analysis, W_{core} and H_{core} are maintained at 1000 nm and 400 nm, respectively.

3. Characterization of RTRR

A photonic ring resonator is a compact and highly sensitive optical device widely used for RI sensing [6]. It operates by coupling light into a circular waveguide where the light resonates at specific wavelengths, which are determined by the optical path length. Changes in the RI of the surrounding medium alter the resonant wavelengths due to shifts in the n_{eff} value of the guided modes. This shift is highly sensitive to even minute changes, enabling the precise detection of biomolecules, gases, or other analytes [4]. The small footprint, compatibility with integrated photonics, and ability to provide real-time, label-free sensing make photonic ring resonators a powerful tool in optical sensing applications [5]. In this section, the fabrication process and details related to optical characterizations are discussed.

3.1. Fabrication Process

The fabrication of PICs involves a series of over a dozen meticulously optimized process steps [31]. In this work, all steps were conducted on 100 mm silicon wafers. The process began with a comprehensive cleaning procedure following standard protocols—RCA SC1, SC2, and Piranha treatments [32]—to remove organic and metallic contaminants from the wafer surface. After pre-cleaning, a wet thermal oxidation process generated a 2.3 µm thick SiO_2 layer, serving as an insulating barrier to separate the silicon substrate from the subsequent Si_3N_4 guiding layer. A 400 nm thick Si_3N_4 layer was then deposited using Low-Pressure Chemical Vapor Deposition (LPCVD). With the base layers in place, the designed pattern was transferred onto the wafer. Given the sub-micron scale of the features, electron beam lithography (EBL) was utilized to achieve high precision and pattern definition.

After exposure and development, Reactive Ion Etching (RIE) was employed to transfer the pattern to the final layer, which was chosen for its anisotropic etching properties and excellent selectivity relative to the lithography resist [4]. Once RIE was complete, the remaining resist was stripped, and a 500 nm thick SiO_2 layer was deposited across the wafer using Plasma-Enhanced Chemical Vapor Deposition (PECVD). At this stage, microfluidic interfaces were integrated into the RTRRs by performing an additional photolithography step followed by localized etching of the SiO_2 layer. The fabrication process concluded with manual cleaving of the wafer to separate the individual chips. The RTRR layout is shown in Figure 5a. Figure 5b presents the microscope image of the RTRR structure, highlighting the localized removal of the SiO_2 layer. The image provides a detailed view of the precise etching and structural integrity of the racetrack, showcasing the effectiveness of the fabrication process in achieving the desired design features.

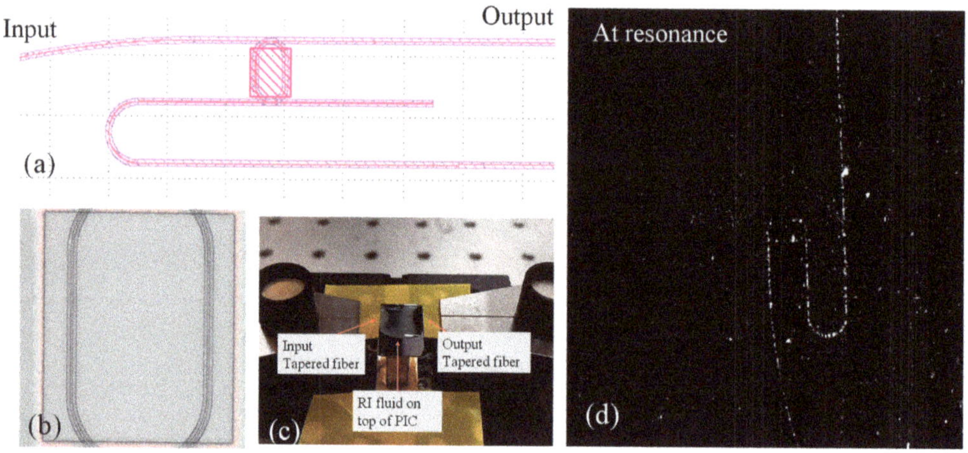

Figure 5. (**a**) Layout of RTRR, (**b**) microscopic image of the RTRR section, (**c**) optical characterization of the PIC, (**d**) image of the ring taken at resonance state with the help of an MIR camera.

3.2. Optical Characterization

A broadband light source (DL-BP1-1501A SLED, Ibsen photonics, Farum, Denmark) was coupled to the bus waveguide using a tapered fiber to ensure efficient light injection. The transmitted light was collected via an output tapered fiber and directed to an optical spectrum analyzer (OSA, Yokogawa AQ6370B, Tokyo, Japan) for characterization. This configuration facilitated accurate measurement of the transmission spectrum of the RTRR. To evaluate the sensitivity of the fabricated structures, certified RI liquids from Cargille Laboratories, Cedar Grove, NJ, USA (Series AAA) were utilized [33]. Renowned for their precision and reliability, these high-quality standard liquids are extensively used in sensing and quality control applications. During the experiment, droplets of RI liquids were carefully deposited onto the sample surface, ensuring uniform coverage for accurate measurements.

After each measurement, the sample was thoroughly cleaned with isopropyl alcohol to eliminate any residual traces, maintaining the integrity of the experimental setup and ensuring repeatability. Figure 5c illustrates the PIC covered with an RI liquid selectively applied to contact only the RTRR. The PIC is positioned on a sample holder, where tapered input and output fibers are precisely aligned and coupled to opposite sides of the chip. This arrangement enables efficient light injection into the bus waveguide and collection of the transmitted light, facilitating the characterization of the device under the influence of the RI liquid. The image of the RTRR in an on-resonance state is taken with the help of an MIR camera (Xenics Xeva), which detects an enhanced light intensity in the ring as a bright spot as shown in Figure 5d.

4. Discussion

The sensitivity of an RI sensor based on a ring resonator is commonly expressed in terms of nanometers per refractive index unit (nm/RIU) [34]. This sensitivity quantifies the shift in the resonance wavelength of the ring resonator in response to changes in the RI of the surrounding medium. The sensitivity of RTRR is calculated by utilizing Equation (5):

$$S_{RTRR} = \frac{\Delta\lambda}{\Delta n} \quad (5)$$

where S_{RTRR} is the sensitivity (nm/RIU) of an RTRR, $\Delta\lambda$ is the shift in the resonance wavelength, and Δn is the change in the RI of the surrounding medium. The transmission spectrum of the RTRR is plotted over the wavelength range of 1520 nm to 1600 nm for various RI fluids (1.414, 1.418, 1.422, and 1.426), as shown in Figure 6a. The resonance wavelength of the RTRR exhibits redshift as the RI of the ambient medium increases. This behavior can be attributed to the dependence of the resonator's mode on the RI of the surrounding medium. As the RI of the ambient medium increases, the n_{eff} of the resonator also increases, which reduces the propagation velocity of light within the resonator. Consequently, to satisfy the resonance condition, the resonant wavelength must increase, resulting in a redshift. Moreover, Figure 6a illustrates six regions marked by green dotted boxes, which highlight the specific wavelength ranges where S_{RTRR} is calculated. These regions offer valuable insight into how the device's sensitivity varies across different wavelengths, showcasing the wavelength-dependent nature of the response. To provide greater clarity, a magnified view of the transmission spectrum is plotted within the wavelength range of 1554 nm to 1576 nm. This detailed representation highlights the prominent resonance dips, which exhibit a redshift as the ambient RI increases, as illustrated in Figure 6b.

As observed in Figure 4b, S_w increases with the operational wavelength due to the enhanced EFR. To further investigate this behavior, we evaluated S_{RTRR} over six distinct wavelength ranges as indicated in Figure 6a. The S_{RTRR} is calculated to be 116.3 nm/RIU, 121.3 nm/RIU, 126.9 nm/RIU, 127.7 nm/RIU, 134.2 nm/RIU, and 143.3 nm/RIU for the respective wavelength regions of 1523–1525 nm, 1530–1532 nm, 1544–1547 nm, 1554–1556 nm, 1574–1576 nm, and 1594–1597 nm, as shown in Figure 6c. This demonstrates that the RTRR's performance can be effectively tailored by operating the device within different wavelength regions, providing versatile control over its functionality.

The quality factor (Q-factor) of a ring resonator is a fundamental metric that characterizes the sharpness of its resonance and its ability to store optical energy [35,36]. It is defined as the ratio of the resonant wavelength to the full width at half maximum (FWHM) of the resonance dip (Equation (6)).

$$Q - \text{factor} = \frac{\lambda_{res}}{\text{FWHM}} \quad (6)$$

where λ_{res} is the resonance wavelength and FWHM is the full width at half maximum.

A higher Q-factor indicates narrower resonance peaks, signifying lower energy losses and enhanced light confinement within the resonator. This attribute is vital for applications such as optical filtering, sensing, and nonlinear optics, where precise control over light–matter interaction is essential. The Q-factor is influenced by intrinsic material properties, waveguide geometry, coupling efficiency, and fabrication quality with higher values often requiring minimized scattering and absorption losses. The Q-factor of the RTRR is calculated for all six regions as indicated in Figure 6a for the RI values of 1.414, 1.418, 1.422, and 1.426, as shown in Figure 6d. The maximum Q-factors achieved are 7701 for region (IV) and 3784.7 for region (V), highlighting the superior resonance characteristics in these wavelength regions. A comprehensive overview of the sensor's performance, including sensitivity, FWHM, and Q-factor values, is provided in Table 2 for detailed reference.

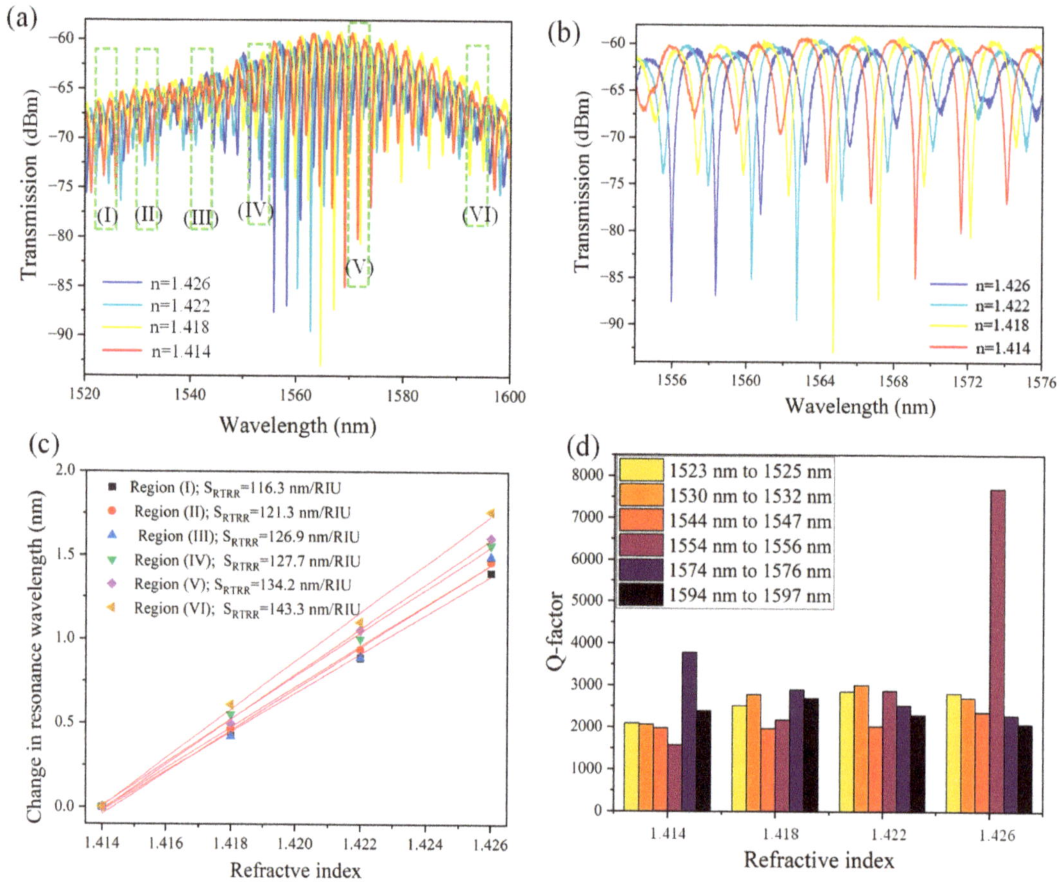

Figure 6. (**a**) Transmission spectrum of an RTRR in the wavelength range of 1520 nm and 1600 nm. The dotted rectangle shows the wavelength regions where sensitivity is calculated, (**b**) the transmission spectrum of an RTRR in the wavelength range IV and V, (**c**) change in resonance wavelength versus refractive index for all six wavelength ranges, (**d**) the Q-factor of RTRR.

Table 2. Sensitivity and Q-factor of RTRR versus wavelength range.

Wavelength range (nm)	1523–1525	1530–1532	1544–1547	1554–1556	1574–1576	1594–1597
Sensitivity (nm/RIU)	116.3	121.3	126.9	127.7	134.2	143.3
FWHM (nm)	0.534	0.508	0.654	0.2	0.415	0.59
Max Q-factor	2854.9	3011.1	2363.5	7701	3784.7	2681.8

5. Concluding Remarks

In conclusion, this paper provides a comprehensive investigation into the wavelength-dependent sensitivity of a silicon nitride (Si_3N_4)-based ridge waveguide, combining detailed numerical analysis with experimental validation. The numerical study is conducted using the finite element method (FEM), which is a powerful computational technique that enables the precise analysis of complex structures and their behavior under various conditions. This approach allows for accurate modeling of the waveguide and its interactions with light, providing valuable insights into the performance of the system. The study begins with an in-depth examination of the modal characteristics of the Si_3N_4 ridge waveguide, focusing on key parameters such as the effective refractive index (n_{eff}), evanescent field ratio (EFR), and propagation losses (α_{prop}). These parameters are pivotal for understanding the interaction between the guided light and the surrounding medium, enabling the optimization of waveguide design for high-performance sensing applications.

The second part of the study experimentally evaluates the wavelength-dependent sensitivity of a racetrack ring resonator (RTRR) fabricated on the Si_3N_4 platform. The results reveal a significant enhancement in sensitivity, increasing from 116.3 nm/RIU to 143.3 nm/RIU as the operating wavelength shifts from 1520 nm to 1600 nm. Additionally, a maximum Q-factor ranging from 3784.7 to 7701 is achieved within the wavelength range of 1554 nm to 1576 nm, demonstrating the RTRR's capability for sharp and precise resonance detection. This observed trend underscores the potential of Si_3N_4-based RTRRs for applications requiring high sensitivity, particularly in the longer wavelength regions. These findings not only validate the effectiveness of Si_3N_4 waveguide designs for advanced sensing applications but also provide valuable insights for the further development of wavelength-dependent photonic devices.

Author Contributions: Conceptualization, M.A.B. and L.K.; methodology, M.A.B.; software, M.A.B.; validation, M.A.B.; formal analysis, M.A.B., L.K.; investigation, M.A.B., L.K., M.S., M.J., D.D., M.F., M.G., B.S.; resources, R.P.; data curation, M.A.B.; writing—original draft preparation, M.A.B.; writing—review and editing, M.A.B., L.K., M.S., M.J.; visualization, M.A.B., L.K., M.S., M.J., M.D., R.P.; supervision, M.A.B.; project administration, M.A.B.; funding acquisition, R.P. All authors have read and agreed to the published version of the manuscript.

Funding: This research received no external funding.

Data Availability Statement: The data will be provided at a reasonable request to the author.

Acknowledgments: The authors acknowledge the constant support of the Warsaw University of Technology in the completion of this work.

Conflicts of Interest: The authors declare no conflicts of interest.

References

1. Blumenthal, D.J.; Heideman, R.; Geuzebroek, D.; Leinse, A.; Roeloffzen, C. Silicon Nitride in Silicon Photonics. *Proc. IEEE* **2018**, *106*, 2209–2231. [CrossRef]
2. Bose, D.; Harrington, M.W.; Isichenko, A.; Liu, K.; Wang, J.; Chauhan, N.; Newman, Z.L.; Blumenthal, D.J. Anneal-Free Ultra-Low Loss Silicon Nitride Integrated Photonics. *Light Sci. Appl.* **2024**, *13*, 156. [CrossRef] [PubMed]
3. Beliaev, L.Y.; Shkondin, E.; Lavrinenko, A.V.; Takayama, O. Optical, Structural and Composition Properties of Silicon Nitride Films Deposited by Reactive Radio-Frequency Sputtering, Low Pressure and Plasma-Enhanced Chemical Vapor Deposition. *Thin Solid Films* **2022**, *763*, 139568. [CrossRef]
4. Butt, M.A.; Kozłowski, Ł.; Golas, M.; Slowikowski, M.; Filipiak, M.; Juchniewicz, M.; Bieniek-Kaczorek, A.; Dudek, M.; Piramidowicz, R. Numerical and Experimental Demonstration of a Silicon Nitride-Based Ring Resonator Structure for Refractive Index Sensing. *Appl. Sci.* **2024**, *14*, 6082. [CrossRef]
5. Castelló-Pedrero, L.; Gómez-Gómez, M.I.; García-Rupérez, J.; Griol, A.; Martínez, A. Performance Improvement of a Silicon Nitride Ring Resonator Biosensor Operated in the TM Mode at 1310 Nm. *Biomed. Opt. Express* **2021**, *12*, 7244–7260. [CrossRef]
6. Hong, Y.; Ge, H.; Hong, J. Compact Biosensors Based on Thin Film Silicon Nitride Microring Resonators. *J. Phys. Conf. Ser.* **2021**, *2012*, 012037. [CrossRef]
7. Wilmart, Q.; El Dirani, H.; Tyler, N.; Fowler, D.; Malhouitre, S.; Garcia, S.; Casale, M.; Kerdiles, S.; Hassan, K.; Monat, C.; et al. A Versatile Silicon-Silicon Nitride Photonics Platform for Enhanced Functionalities and Applications. *Appl. Sci.* **2019**, *9*, 255. [CrossRef]

8. Ji, X.; Okawachi, Y.; Gil-Molina, A.; Corato-Zanarella, M.; Roberts, S.; Gaeta, A.L.; Lipson, M. Ultra-Low-Loss Silicon Nitride Photonics Based on Deposited Films Compatible with Foundries. *Laser Photonics Rev.* **2023**, *17*, 2200544. [CrossRef]
9. Lin, Y.; Yong, Z.; Luo, X.; Azadeh, S.S.; Mikkelsen, J.C.; Sharma, A.; Chen, H.; Mak, J.C.C.; Lo, P.G.-Q.; Sacher, W.D.; et al. Monolithically Integrated, Broadband, High-Efficiency Silicon Nitride-on-Silicon Waveguide Photodetectors in a Visible-Light Integrated Photonics Platform. *Nat. Commun.* **2022**, *13*, 6362. [CrossRef]
10. Romero-García, S.; Merget, F.; Zhong, F.; Finkelstein, H.; Witzens, J. Silicon Nitride CMOS-Compatible Platform for Integrated Photonics Applications at Visible Wavelengths. *Opt. Express* **2013**, *21*, 14036–14046. [CrossRef]
11. Liu, J.; Huang, G.; Wang, R.N.; He, J.; Raja, A.S.; Liu, T.; Engelsen, N.J.; Kippenberg, T.J. High-Yield, Wafer-Scale Fabrication of Ultralow-Loss, Dispersion-Engineered Silicon Nitride Photonic Circuits. *Nat. Commun.* **2021**, *12*, 2236. [CrossRef]
12. Dow, H.S.; Kim, W.S.; Lee, J.W. Thermal and Electrical Properties of Silicon Nitride Substrates. *AIP Adv.* **2017**, *7*, 095022. [CrossRef]
13. Pezzotti, G.; McEntire, B.J. The Future of Silicon Nitride: Biomedical Applications and Beyond. In *Silicon Nitride Bioceramics*; Bal, B.S., McEntire, B.J., Pezzotti, G., Eds.; Springer International Publishing: Cham, Switerland, 2024; pp. 415–431. ISBN 978-3-031-67047-3.
14. Krstic, Z.; Krstic, V.D. Silicon Nitride: The Engineering Material of the Future. *J. Mater. Sci.* **2012**, *47*, 535–552. [CrossRef]
15. Afroozeh, A. Analysis of Optical Modulator Based on Silicon Waveguide Using FDTD. *Silicon* **2022**, *14*, 839–849. [CrossRef]
16. Butt, M.A.; Janaszek, B.; Piramidowicz, R. Lighting the Way Forward: The Bright Future of Photonic Integrated Circuits. *Sens. Int.* **2025**, *6*, 100326. [CrossRef]
17. Barrios, C.A. Optical Slot-Waveguide Based Biochemical Sensors. *Sensors* **2009**, *9*, 4751–4765. [CrossRef]
18. Butt, M.A.; Kaźmierczak, A.; Tyszkiewicz, C.; Karasiński, P.; Piramidowicz, R. Mode Sensitivity Exploration of Silica–Titania Waveguide for Refractive Index Sensing Applications. *Sensors* **2021**, *21*, 7452. [CrossRef]
19. Lin, F.-L.C. Modal Characteristics of Crossed Rectangular Waveguides. *IEEE Trans. Microw. Theory Tech.* **1977**, *25*, 756–763. [CrossRef]
20. Lai, C.-H.; You, B.; Lu, J.-Y.; Liu, T.-A.; Peng, J.-L.; Sun, C.-K.; Chang, H. Modal Characteristics of Antiresonant Reflecting Pipe Waveguides for Terahertz Waveguiding. *Opt. Express* **2010**, *18*, 309–322. [CrossRef]
21. Kovacs, B.; Kraft, F.A.; Szabo, Z.; Nazirizadeh, Y.; Gerken, M.; Horvath, R. Near Cut-off Wavelength Operation of Resonant Waveguide Grating Biosensors. *Sci. Rep.* **2021**, *11*, 13091. [CrossRef]
22. Cut-off Wavelength Measurements of Ti:LiNbO3 Channel Waveguides-IOPscience. Available online: https://iopscience.iop.org/article/10.1088/0957-0233/5/9/014 (accessed on 26 December 2024).
23. Butt, M.A. Dielectric Waveguide-Based Sensors with Enhanced Evanescent Field: Unveiling the Dynamic Interaction with the Ambient Medium for Biosensing and Gas-Sensing Applications—A Review. *Photonics* **2024**, *11*, 198. [CrossRef]
24. Isaacs, S.; Hajoj, A.; Abutoama, M.; Kozlovsky, A.; Golan, E.; Abdulhalim, I. Resonant Grating without a Planar Waveguide Layer as a Refractive Index Sensor. *Sensors* **2019**, *19*, 3003. [CrossRef] [PubMed]
25. Odeh, M.; Twayana, K.; Sloyan, K.; Villegas, J.E.; Chandran, S.; Dahlem, M.S. Mode Sensitivity Analysis of Subwavelength Grating Slot Waveguides. *IEEE Photonics J.* **2019**, *11*, 1–10. [CrossRef]
26. Hunsperger, R.G. Losses in Optical Waveguides. In *Integrated Optics: Theory and Technology*; Hunsperger, R.G., Ed.; Springer: Berlin/Heidelberg, Germany, 2002; pp. 93–111. ISBN 978-3-540-38843-2.
27. Luke, K.; Okawachi, Y.; Lamont, M.R.E.; Gaeta, A.L.; Lipson, M. Broadband Mid-Infrared Frequency Comb Generation in a Si_3N_4 Microresonator. *Opt. Lett.* **2015**, *40*, 4823–4826. [CrossRef]
28. Kischkat, J.; Peters, S.; Gruska, B.; Semtsiv, M.; Chashnikova, M.; Klinkmüller, M.; Fedosenko, O.; Machulik, S.; Aleksandrova, A.; Monastyrskyi, G.; et al. Mid-Infrared Optical Properties of Thin Films of Aluminum Oxide, Titanium Dioxide, Silicon Dioxide, Aluminum Nitride, and Silicon Nitride. *Appl. Opt.* **2012**, *51*, 6789–6798. [CrossRef]
29. COMSOL: Multiphysics Software for Optimizing Designs. Available online: https://www.comsol.com/ (accessed on 26 September 2023).
30. Chandra, V.; Ranjan, R. Performance Analysis of Different Slot Waveguide Structures for Evanescent Field Based Gas Sensor Applications. *Opt. Quantum Electron.* **2021**, *53*, 457. [CrossRef]
31. Butt, M.A. Integrated Optics: Platforms and Fabrication Methods. *Encyclopedia* **2023**, *3*, 824–838. [CrossRef]
32. Du, C.; Zhao, Y.; Li, Y. Effect of Surface Cleaning Process on the Wafer Bonding of Silicon and Pyrex Glass. *J. Inorg. Organomet. Polym. Mater.* **2023**, *33*, 673–679. [CrossRef]
33. Labs, C. Refractive Index Liquids—Cargille Labs. Available online: https://www.cargille.com/ (accessed on 12 November 2024).
34. Butt, M.A.; Kozłowski, Ł.; Piramidowicz, R. Optimized Hybrid Plasmonic Waveguide-Based Ring Resonator for Advanced Refractive Index Sensing. *J. Opt.* **2024**, *26*, 075802. [CrossRef]

35. Zhang, X.; Zhou, C.; Luo, Y.; Yang, Z.; Zhang, W.; Li, L.; Xu, P.; Zhang, P.; Xu, T. High Q-Factor, Ultrasensitivity Slot Microring Resonator Sensor Based on Chalcogenide Glasses. *Opt. Express* **2022**, *30*, 3866–3875. [CrossRef]
36. Cicek, K.; Boybay, M.S.; Topkaya, R.; Agarwal, A.M. C-Band Operating Plasmonic Sensor with a High Q-Factor/Figure of Merit Based on a Silicon Nano-Ring. *Appl. Opt.* **2023**, *62*, 1290–1295. [CrossRef]

micromachines

MDPI

Review

A Review of Transverse Mode Adaptive Control Based on Photonic Lanterns

Yao Lu [1,2,*], Zongfu Jiang [3], Zilun Chen [3], Zhuruixiang Sun [1] and Tong Liu [4,*]

1 Space Engineering University, Beijing 101416, China
2 National Key Laboratory of Space Target Awareness, Beijing 101416, China
3 College of Advanced Interdisciplinary Studies, National University of Defense Technology, Changsha 410073, China
4 Department of Aerospace Science and Technology, Space Engineering University, Beijing 101416, China
* Correspondence: lybenben1@163.com (Y.L.); liutong719@163.com (T.L.)

Abstract

With the widespread application of fiber laser technology in industries, communications, medical fields, and beyond, the demand for controlling the spatial modes of their output beams has been increasingly growing. Traditional mode control methods are constrained by factors such as device power thresholds, system complexity, and cost, making it difficult to meet the requirements for high-power, high-purity, and rapidly switchable multimode regulation. This paper reviews adaptive mode control technology based on photonic lanterns (PLs). By integrating ideas from adaptive optics and photonics, this technology utilizes photonic lanterns to achieve efficient mode evolution from single-mode to multimode fibers. Combined with optimization algorithms, it enables real-time regulation of input phases, thereby producing stable, high-purity target modes or mode superposition fields at the multimode output end. The paper systematically introduces the structural classifications, propagation characteristics, and fabrication processes of photonic lanterns, as well as the mode evolution mechanisms in different types of photonic lanterns. It elaborates in detail on the structural design, algorithm implementation, and experimental validation of the adaptive control system based on photonic lanterns. Furthermore, it explores the application prospects of this technology in areas such as suppressing transverse mode instability, mode-division multiplexing communications, particle manipulation, and high-resolution spectral measurements. The results demonstrate that the all-fiber adaptive mode control system based on photonic lanterns offers advantages such as compact structure, low loss, fast response, and strong scalability.

Keywords: photonic lantern; mode control; fiber laser

1. Introduction

T.H. Maiman created the first solid-state laser in history in 1960. Snitzer proposed the use of fiber to produce laser light the next year, and in 1964, he successfully demonstrated the first fiber laser. At the same time that rare-earth ion-doped low-loss fibers and double-cladding fibers were designed and developed, semiconductor laser pumping technology advanced significantly in the 1980s. Over the subsequent decades, fiber laser performance has made tremendous strides, with fiber lasers finding excellent applications in industrial, defense, medical, scientific, and entertainment fields.

As research into fiber lasers has deepened, it has become increasingly recognized that beyond power, wavelength, and polarization, the spatial distribution characteristics of fiber lasers, particularly transverse mode characteristics, are equally important factors determining the practical application value of fiber lasers. Transverse modes are defined as stable field distributions existing in the transverse plane perpendicular to the propagation direction. Different transverse modes possess different application potential: for example, fundamental modes with excellent beam quality and Gaussian intensity distribution can be applied to high-energy laser systems for industrial production and defense applications [1]; OAM modes with helical phase distribution and carrying orbital angular momentum can realize particle manipulation and super-resolution imaging [2,3]. In addition, orthogonality between modes allows them to be multiplexed or coded as new data stream carriers to increase the capacity and efficiency of the communication system [4].

Therefore, how to obtain a stable mode field (or mode superposition field) on demand and how to realize mode control efficiently have become popular research directions in the field of fiber lasers.

Current fiber laser mode control schemes mainly originate from adaptive optics and photonics concepts.

Adaptive optics-based schemes primarily include the following:

(1) adaptive wave-front correction performed on the single aperture fiber laser output, with the modulator selected from Deformable Mirrors [5], liquid crystal spatial light modulators (LC-SLM) [6], polarization controllers, or lensless on-chip microscopy arrays [7] and other devices; (2) adaptive control of the coherent beam combination from fiber laser arrays [8], controlling the phase, intensity, polarization, tilt, and other parameters of each aperture output beam; and (3) enhancement of the loss of non-target modes through adaptive control of the multi-point bending and squeezing [9] of fibers, thereby selecting target modes.

Photonics-based mode control schemes primarily include the following:

(1) designing specialty fibers (such as tapered fibers [10], chirally-coupled-core fibers [11], photonic crystal fibers [12], leakage channel fibers [13], photonic bandgap fibers [14], multi-trench rod-type fibers [15], etc.) to achieve intra-fiber mode filtering; (2) designing long-period fiber gratings to achieve intra-fiber mode conversion [16]; (3) designing fiber couplers [17] to achieve intra-fiber mode selection; and (4) designing metasurface devices/chips at fiber output facets [18] or within fiber laser cavities to achieve mode conversion.

However, all of these mode control schemes are limited by one or more factors, including but not limited to the following: low power threshold of key devices, large system losses, difficult thermal management, high process costs, and system complexity. Facing practical application demands for higher power, higher mode purity, faster mode switching speeds, and more tunable mode numbers, there is an urgent need to find a new implementation pathway for fiber laser mode control. Photonic lantern-based mode adaptive control technology has emerged in response to this need [19]. This technology simultaneously integrates both adaptive optics and photonics concepts, with the core device—the photonic lantern—serving as a novel all-fiber "mode converter" [20] that enables efficient evolution of fundamental mode beams in single-mode fiber (SMF) bundles into multimode beams in a single multimode fiber (MMF). Combined with adaptive optics technology for active control of phases at the single-mode end, stable and controllable mode fields (or mode superposition fields) can ultimately be output at the multimode end.

2. Mode Converter: Photonic Lantern

Photonics is defined as the science of shaping the flow of light beams. Applications for photonics have covered various light operations, including light transmission, filtering, noise suppression, interferometry, and more. Common photonic devices include fiber Bragg gratings, fiber couplers, optical circulators, beam shapers, optical switches, etc. However, the performance of these devices often only works effectively in single-mode fiber systems [21].

Driven by application demands in the astronomical field, in 2005, J. Bland-Hawthorn proposed the idea of "using multimode fibers to achieve single-mode fiber Bragg grating filtering performance" during a visit to the University of Bath. Inspired by this, S. G. Leon-Saval, T. A. Birks, and others designed a novel all-fiber mode converter—the photonic lantern (PL) [22]. As shown in the schematic diagram in Figure 1a, it achieves light coupling and evolution through a gradual taper transition region between one multimode fiber (MMF) and multiple single-mode fibers (SMFs), realizing bidirectional lossless transmission between MMF and SMFs. Simultaneously, photonic devices such as fiber Bragg gratings can be inscribed on the single-mode fiber bundle to achieve filtering, providing the possibility for various photonics-based light operations in multimode fiber systems. For SMFs, since only one mode exists, the optical energy is highly concentrated at the core center. When light is reflected by the grating, the photothermal heating is also localized in this central region. This symmetric and concentrated heat source enables highly uniform and efficient radial heat conduction to the cladding and coating layers, ultimately dissipating into the surrounding environment, which facilitates efficient thermal management. Similarly, integrating other photonic devices onto SMFs is feasible, ensuring that the power threshold of photonic lanterns can be significantly high. Due to the structural appearance of this mode converter being similar to decorative lanterns used for illumination in daily life (as shown in Figure 1b), it was named the "photonic lantern." In subsequent specific research and applications, it has become customary to refer to one MMF-SMFs transition structure as one photonic lantern (PL), and an entire symmetric MMF-SMFs-MMF structure as a photonic lantern pair (PL-pair), as shown in Figure 1c.

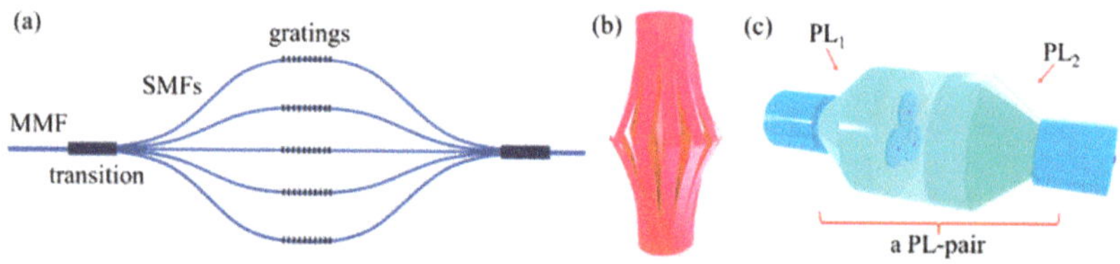

Figure 1. Schematic diagram of photonic lantern structure. (**a**) schematic diagram of the PL; (**b**) paper lantern in daily life; (**c**) a PL-pair.

As pioneers in the photonic lantern field, T. A. Birks from the University of Bath and S. G. Leon-Saval from the University of Sydney not only first provided the structural design of photonic lanterns but also first conducted detailed analysis of their light transmission characteristics. In 2010, they calculated the propagation constants of modes supported by various cross-sections of photonic lanterns based on the beam propagation method (BPM), revealing the most fundamental principles of the mode evolution process [23]. In 2012, Nicolas K. Fontaine and others from Bell Laboratories studied the influence of different geometric arrangements of single-mode fiber bundles on photonic lantern performance and first introduced transmission matrices to describe the light transmission characteristics of photonic lanterns [24]. To more comprehensively analyze the polarization characteristics of light transmission in photonic lanterns, in 2021, S. G. Leon-Saval led his team to calculate the vector transmission matrix of photonic lanterns using nonlinear neural networks [25]. Thus, the theoretical framework for photonic lanterns was essentially established.

Taking light transmission from the SMFs end to the MMF end as an example, as shown in Figure 2, the specific process of mode evolution is as follows: at one end of the SMFs bundle (point A), light beams exist as fundamental modes with small mode fields propagating independently in separated cores; after entering the taper region (points B, C, and D), as the cores continuously become thinner and closer together, they form a multi-core waveguide structure, causing the light beams to evolve into supermode forms; and at the end of the taper region (points E, F), the modes supported by sufficiently thin and close core groups are essentially the same as the large mode field guided modes in the MMF. This ensures that when light enters the MMF, the mode field overlap integral to both ends of the fusion splice is high, and the light ultimately efficiently evolves into large mode field MMF guided modes or guided mode superposition fields (points G and H).

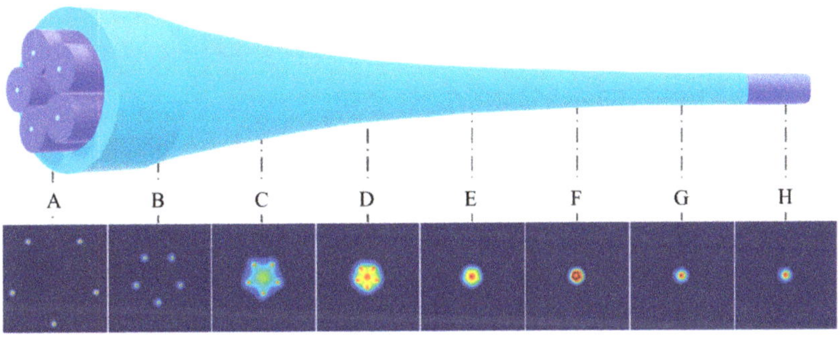

Figure 2. Mode evolution process in a 5 × 1 photonic lantern with equal amplitude and zero phase difference input.

According to the criterion of whether the single-channel mode evolution results of photonic lanterns exhibit differences, photonic lanterns are divided into two major categories [26]: non-mode-selective photonic lanterns (or mode-group-selective photonic lanterns) and mode-selective photonic lanterns. Non-mode-selective photonic lanterns consist of single-mode fiber bundles composed of completely identical single-mode fibers. When light enters this type of photonic lantern from different signal channels, the mode evolution process exhibits good symmetry, and the output light field obtained at the multimode end is generally a superposition field containing multiple modes. Mode-selective photonic lanterns require the use of single-mode fibers with different structural parameters for bundling to ensure that the fundamental modes supported by each single-mode signal channel have different propagation constants [27], thereby causing differential mode evolution and ultimately obtaining different evolution results. Through rigorous design of single-mode fiber core size gradient distributions and the overall geometric structure design of photonic lanterns, this type of photonic lantern can achieve single-channel light transmission with mode evolution output light fields that are high-purity, large mode field single modes.

Both types of photonic lanterns have their respective advantages: mode-selective photonic lanterns can achieve mode selection without relying on any external control means, requiring only changing the input channel for mode switching; non-mode-selective photonic lanterns, due to their single-mode fiber bundles using the same fiber specifications, have advantages mainly reflected in the following aspects: (1) strong channel number scalability, as non-mode-selective photonic lanterns do not require special core diameter size gradient distributions for single-mode fiber bundles; (2) low fabrication process difficulty, as single-mode fiber bundle assembly is easy, generally achieved by arranging identical single-mode fibers in layered close-packed regular polygonal arrangements; and (3) when working with control systems, non-mode-selective photonic lanterns, due to their good symmetry, can make control algorithm convergence faster and precisely achieve control of mode superposition fields.

According to waveguide theory, the single-mode fiber core bundle taper region of photonic lanterns is an irregular waveguide with longitudinally non-uniform refractive index distribution. The main characteristic of irregular waveguides is that strictly defined modes do not exist, i.e., there are no analytical solutions in the form of $e(x,y)\cdot e\hat{}(-i\beta z)$. In this case, we can find a regular waveguide segment such that the field within the irregular waveguide can be expanded as a series sum of modes of this regular waveguide. This expansion includes not only the discrete sum of guided modes but also the continuous sum of radiation modes (where radiation modes are radial standing wave distributions formed by the superposition of incident and reflected light outside the fiber core).

The focus of studying photonic lantern mode evolution processes is the calculation of eigenmodes including guided modes and radiation modes, and the calculation of mode superposition and coupling between them. Methods such as the beam propagation method (BPM) [28] or mode expansion propagation method [29] can be used to numerically simulate the light propagation process in photonic lanterns, obtaining the one-to-one correspondence relationship between input and output light fields of photonic lanterns. Mathematically, the relationship between input and output light fields of photonic lanterns can be compactly expressed as follows:

$$u = M \cdot v, \tag{1}$$

where M is the transmission matrix of the photonic lantern, v is a column vector describing the light field information at the single-mode end of the photonic lantern, $v^T = [v_1e^{i\theta}{}_1, v_2e^{i\theta}{}_2, v_3e^{i\theta}{}_3, \ldots, v_je^{i\theta_j}, \ldots]$ (where v_j and θ_j represent the amplitude and phase of the fundamental mode light field input from the j-th single-mode fiber, respectively), and u is a column vector describing the light field information at the multimode end of the photonic lantern, $u^T = [u_1e^{i\varphi}{}_1, u_2e^{i\varphi}{}_2, u_3e^{i\varphi}{}_3, \ldots, u_ke^{i\varphi k}, \ldots]$ (where u_k and φ_k represent the amplitude and phase of the k-th eigenmode in the output superposition field, respectively).

For example, when using linearly polarized mode (LP mode) sequences as eigenmode sequences, the transmission matrix of an ideal 3×1 mode-selective photonic lantern is Equation (2), while that of a 3×1 non-mode-selective photonic lantern is Equation (3):

$$M^{\text{ideal}}_{\text{LP},3\times1\text{MSPL}} = \begin{bmatrix} 1 & 0 & 0 \\ 0 & 0 & 1 \cdot e^{i(0.2\pi)} \\ 0 & 1 \cdot e^{i(-0.4\pi)} & 0 \end{bmatrix}, \tag{2}$$

$$M^{\text{ideal}}_{\text{LP},3\times1\text{MGPL}} = \begin{bmatrix} \frac{1}{\sqrt{3}} & \frac{1}{\sqrt{3}} & \frac{1}{\sqrt{3}} \\ \frac{1}{\sqrt{6}} & \frac{1}{\sqrt{6}} & \frac{2}{\sqrt{6}}e^{i\pi} \\ \frac{1}{\sqrt{2}} & \frac{1}{\sqrt{2}}e^{-i\pi} & 0 \end{bmatrix}, \tag{3}$$

When the taper region of the photonic lantern is sufficiently long and satisfies the adiabatic approximation condition, the photonic lantern can be viewed as a linear optical device, and its transmission matrix is invertible. Therefore, to obtain a specific mode at the multimode fiber output end of the photonic lantern, one only needs to set the corresponding term for that mode in column vector u to 1 and other terms to 0, and then calculate the required input light field combination according to Equation (1):

$$v_{\text{target mode}} = M^{-1} \cdot u_{\text{target mode}}, \tag{4}$$

As the core device for achieving mode evolution between single-mode fiber systems and multimode fiber systems, the performance and reliability of photonic lanterns greatly affect mode control effectiveness. Therefore, many researchers are dedicated to continuously optimizing photonic lantern fabrication processes. T. A. Birks categorized photonic lantern fabrication schemes into three types: direct ultrafast laser inscription, tapering multicore fiber, and tapering SMF-tube bundle method. Among these, the tapering SMF-tube bundle method is most favored.

Taking a three-channel example, the structural side view of a tapering SMF-tube bundle photonic lantern is shown in Figure 3. Its fabrication steps are as follows: first, a group of separated single-mode fiber bundles are closely arranged and inserted into a low-index glass tube, then simultaneously fused and tapered. As the transverse dimensions decrease, the fiber cores in the single-mode fiber bundle continuously approach each other and become thinner, forming the core of a new waveguide structure, while the low-index glass tube becomes the new cladding. Second, the tube-single-mode fiber bundle is cut at the waist region (the area with the smallest diameter). Finally, the waist region end face of the tube-single-mode fiber bundle is fusion-spliced with a matching multimode fiber. This method produces photonic lanterns with separated single-mode end fibers, making it easy to achieve light field control at the single-mode end [30]. Additionally, the raw materials for making such photonic lanterns are common step-index fibers, ensuring good compatibility with fiber laser seed sources, amplifier stages, and other fiber systems.

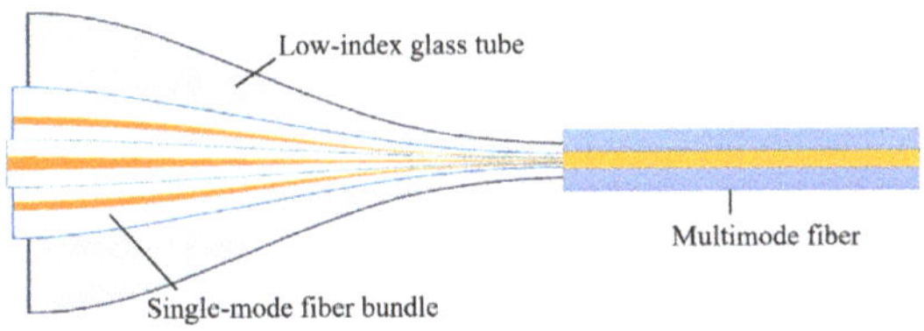

Figure 3. Side view of the structure of the photonic lantern fabricated by the tapering SMF-tube bundle method.

Due to practical factors such as small fiber cross-sectional dimensions and reduced waveguide structure strength after fusion tapering, process details have a significant impact on photonic lantern yield and performance. Key process details requiring optimization include the following: device cleanliness, single-mode fiber bundle assembly techniques, fusion tapering parameters, and fusion splice alignment [31]. In 2009, Noordegraaf et al. from Denmark fabricated the first high-efficiency, low-loss seven-channel non-mode-selective photonic lantern based on the tube fusion tapering method [32]. In 2015, Velazquez-Benitez et al. from the University of Central Florida completed the design and fabrication of a six-channel mode-selective photonic lantern using the tube method [27], achieving insertion loss of only 0.2–0.3 dB and mode purity exceeding 9 dB for all channels. In 2021, John J. Davenport et al. from Germany conducted detailed studies on the influence of bundling conditions on photonic lantern performance in tube method fabrication schemes and completed the fabrication and testing of multi-channel (7, 19, and 37 channels) non-mode-selective photonic lanterns [33].

3. Transverse Mode Adaptive Control System

The basic principle of adaptive optics is to compensate the transmitted signal based on the detected received signal, belonging to feedback-type active control systems, roughly divided into two categories: conjugate systems and optimized systems. Conjugate systems require real-time detection of actual input signals and calculation of the difference between them and ideal input signals reverse-calculated from target output signals; then, they compensate for this difference. If photonic lanterns are applied in such systems, it would be necessary to interrupt the input end and introduce a section of free-space optical path to detect input signals, which would undermine the advantages of photonic lanterns as all-fiber devices.

Optimized adaptive control systems calculate a certain evaluation function at the output signal end in real-time to reflect system compensation performance, using optimization algorithms to help the input end search for appropriate modulation signals so that the evaluation function at the output end converges to a unique extremum. Optimized adaptive control systems avoid measurement of input end signals and calculation processes for modulation signals, simplifying experimental optical path complexity while providing better scalability for system channel numbers.

Borrowing from the optimized structure, a photonic lantern-based mode adaptive control system is designed, as shown in Figure 4: a single-frequency linearly polarized seed laser travels along single-mode fibers, passes through a splitter, then through phase modulators (PM) and pre-amplifiers (Pre-AMP) on each single-mode polarization-maintaining line, and is injected into the single-mode fiber end of the photonic lantern, outputting at the multimode fiber end of the photonic lantern. A PM is used for on-line phase modulation of each path under algorithm control, while Pre-AMP is used for off-line adjustment of input power for each path of the photonic lantern.

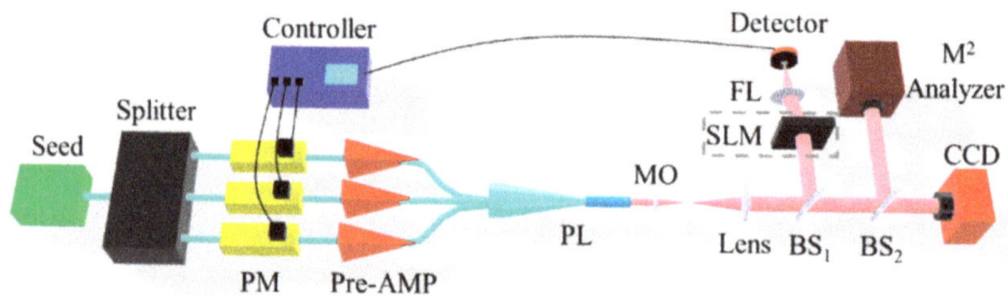

Figure 4. Schematic diagram of mode adaptive control system based on 3 × 1 photonic lantern.

In the external optical path, a microobjective (MO) and a large focal length lens combination form a 4F system (4F Fourier lens system) to magnify the near-field pattern size at the photonic lantern output end. Beam Splitter 1 (BS1) splits out one output signal path to the detector, providing an evaluation function for the controller. When necessary, the output signal can pass through a spatial light modulator (SLM) for optical filtering preprocessing. Beam Splitter 2 (BS2) splits the output beam to a CCD camera and M^2 analyzer for near-field spot detection and beam quality analysis of control effects.

The mode evolution process in photonic lanterns stems from continuous mode superposition and coupling of light within them. The occurrence of these two interactions depends on the coherence between input light fields of each path in the photonic lantern. The fundamental mode beams at each path of the photonic lantern input end shown in Figure 4 are emitted from the same single-frequency seed source, with essentially consistent beam frequencies between paths and long coherence length. Additionally, all devices before the photonic lantern input end are linearly polarized devices connected by polarization-maintaining fibers. Therefore, ensuring essentially the same optical path length between paths can satisfy coherence conditions.

Phase noise commonly exists in practical fiber laser systems, where phase noise introduced by environmental disturbances generally has characteristic frequencies below the kilohertz level, and phase noise introduced by gain media themselves generally has characteristic frequencies below the hundred-hertz level. To ensure that the mode adaptive control system can achieve modulation and locking of input phases for each path, the system's control bandwidth must be higher than the highest characteristic frequency of phase noise.

Lithium niobate ($LiNbO_3$) crystal phase modulators are generally chosen as modulation devices, offering advantages such as high response frequency (GHz level) and half-wave voltage that does not drift with modulation frequency. Additionally, many optimization algorithms can easily generate disturbance signals of several hundred kHz. Therefore, as long as suitable optimization algorithms are found and parameters are tuned to enable fast convergence (within 100 loops) and achieve control bandwidth above the kHz level, phase modulation and locking can be realized.

Commonly used optimization algorithms include the following: dithering algorithm, genetic algorithm, hill climbing algorithm, simulated annealing algorithm, and Stochastic Parallel Gradient Descent (SPGD) algorithm. Their working principles determine the response time for single-step operations and the number of iterations required for convergence, thereby determining control bandwidth. The SPGD algorithm is an efficient algorithm that performs gradient estimation each time modulation signals are updated [34]. The modulation signal iteration formula for this algorithm is as follows: $\boldsymbol{u}^{(m+1)} = \boldsymbol{u}^{(m)} + \gamma\delta\boldsymbol{u}^{(m)}\Delta J^{(m)}$, where m is the current iteration number, γ is the gain coefficient, $\delta\boldsymbol{u}^{(m)}$ is the random disturbance voltage magnitude in the m-th iteration, and ΔJ is the evaluation function change amount, providing gradient information for iteration. The introduction of this parameter reduces the number of iterations required for evaluation function convergence, thereby improving overall system control bandwidth. Additionally, through appropriate evaluation function selection and further parameter optimization, the SPGD algorithm also offers advantages such as strong system scalability and high power fluctuation tolerance.

Without loss of generality, taking a 3×1 non-mode-selective photonic lantern achieving mode control at a 30 μm core diameter multimode fiber output end as an example, numerical simulation of the execution process of the photonic lantern-based mode control system is performed. The evaluation function J is taken as the power in bucket (PIB) of the near-field spot at the multimode output end of the photonic lantern, with the specific position of the "bucket" determined according to different mode control objectives. Through parameter optimization, it is determined that when disturbance voltage $\delta = 0.07$ and gain coefficient $\gamma = 200$, the average convergence speed of the evaluation function is fastest.

Figure 5 shows numerical simulation results for achieving LP_{01} mode or LP_{11o} mode output after executing control algorithms with 50 random fundamental mode combinations as initial inputs. During calculations, initial phases for each path are set as random phases within the 0 to 2π range, and input power for each path at the single-mode fiber end randomly fluctuates by 10% based on ideal values calculated according to Equation (4). Figure 5a shows the change in LP_{01} mode content obtained through real-time mode decomposition of output spots when LP_{01} mode is the control target. In this case, the "bucket" is selected at the spot center (position shown by red dot in the figure). The algorithm basically completes convergence within 50 loop cycles. After convergence, the output light field remains stable, with an average convergence value of LP_{01} mode content of 0.9514 across 50 calculations, minimum value of 0.9251, and maximum value of 0.9798. Figure 5b shows control effects when LP_{11o} mode is the control target. In this case, the "bucket" is selected at a position away from the spot center. The algorithm convergence speed has a slight decrease, basically completing convergence within 60 loop cycles. After convergence, the average convergence value of LP_{11o} mode content is 0.9432.

The above results show that after selecting appropriate control parameters, even under conditions of 10% power fluctuation at each input path, the SPGD algorithm can still efficiently complete phase locking and achieve stable, high-purity target mode output.

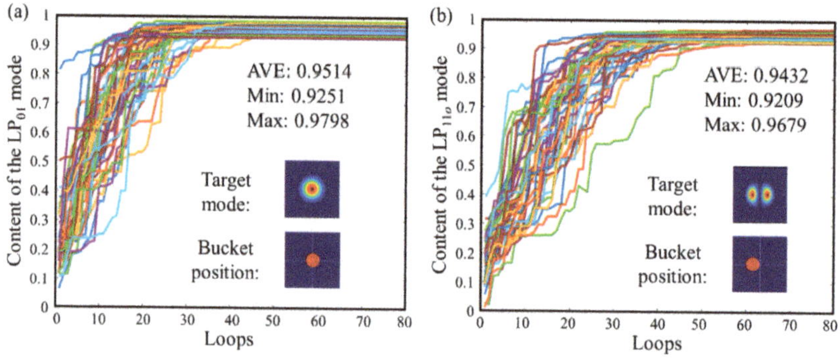

Figure 5. Using the SPGD algorithm, and achieving (**a**) LP_{01} mode and (**b**) LP_{11o} mode control based on a 3 × 1 non-mode-selective photonic lantern.

Building the mode adaptive control system shown in Figure 4, Figure 6 shows the changes in near-field light intensity distribution, beam quality, and oscilloscope-displayed detector voltage signals of measured output beams before and after control algorithm closed-loop operation when the fundamental mode is the control target. When the system is open-loop (SPGD off), the near-field spot morphology continuously changes and is in a multimode superposition state, the M^2 factor continuously fluctuates with poor beam quality, and the detector voltage signal continuously oscillates with low amplitude. When the system is closed-loop (SPGD on), the spot locks to a quasi-fundamental mode form (fundamental mode content exceeding 0.95) in a very short time, the M^2 factor slightly fluctuates below 1.2, and the oscilloscope shows that the voltage signal amplitude is high and remains stable.

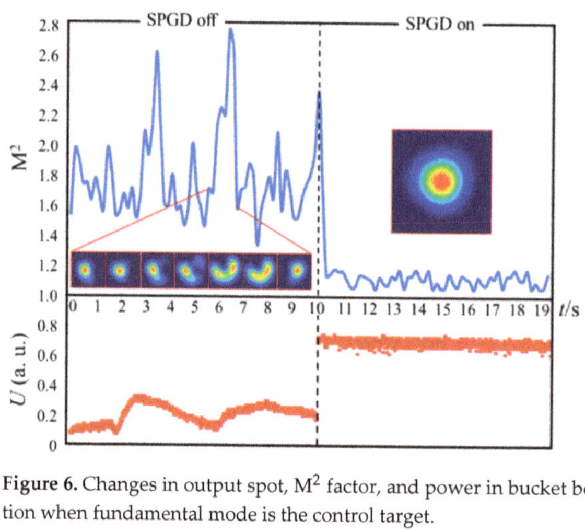

Figure 6. Changes in output spot, M^2 factor, and power in bucket before and after algorithm execution when fundamental mode is the control target.

Figure 7 shows the changes in mode proportion of measured output beams before and after control algorithm closed-loop operation when LP_{11o} mode is the control target. In this case, the relative positions between the bucket and spot centroid and the bucket size are changed by adjusting the positions of the detector and aperture and their distances from the lens. Experimental results show that when the system is open-loop, the mode content of output beams continuously changes with energy conversion between modes, while when the system is closed-loop, the output spot stabilizes in a quasi- LP_{11o} mode form with LP_{11o} mode proportion of approximately 0.924.

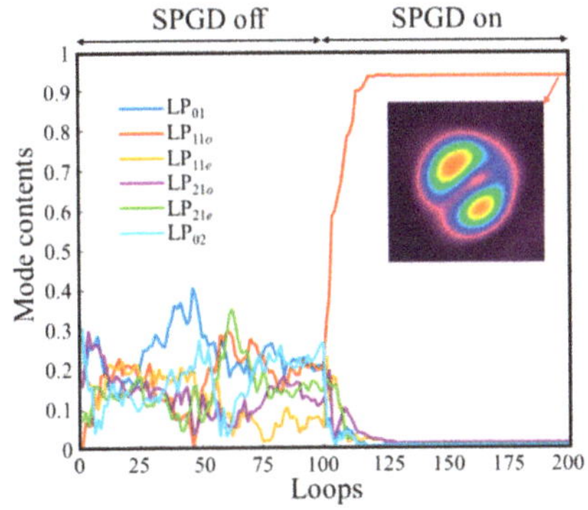

Figure 7. Changes in output beam mode proportion when LP_{11o} mode is the control target and near-field light intensity distribution after closed-loop system operation.

4. Application of All-Fiber Mode Control Systems

Photonic lantern-based mode control systems achieve all-fiber implementation and are expected to meet practical application demands for higher power, higher mode purity, faster mode switching speeds, and more tunable mode numbers. They have broad application prospects in industrial manufacturing, optical fiber communications, and frontier exploration fields, as shown in Figure 8.

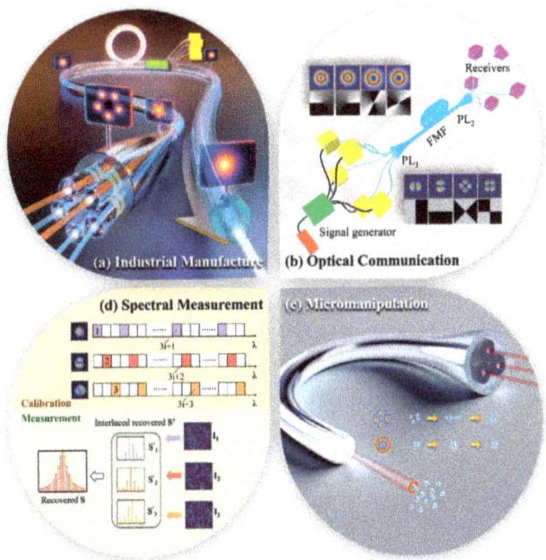

Figure 8. Schematic diagram of application scenarios for photonic lantern-based mode control.

4.1. Industrial Manufacturing

In 2000, Beck et al., to improve laser processing precision, controllable dimensions, and remote operation capabilities, coupled laser light emitted by Nd: YAG lasers with fibers and first used fundamental mode beam output from fiber end faces for material processing [35]. Since then, fiber lasers have gradually developed into indispensable tools in the industrial manufacturing field. According to the basic principles of light-matter interaction, the key for fiber lasers to meet application demands in industrial manufacturing and other high-energy laser fields (such as directed energy technology and gravitational wave detection) is using fiber lasers to emit high-power, mode-stable light beams.

In 2008, Dawson et al. theoretically calculated [36] and predicted that the ultimate output power limit for single-frequency fiber lasers is 1.86 kW, and for broadband fiber lasers is 36.6 kW. However, this prediction was based on the assumption that fiber mode field size is unlimited. To achieve ultimate power output, implementation in 90 μm core diameter or even larger fibers would be required, making it difficult to guarantee output beam mode stability. Experimentally, in 2011, Eidam et al. discovered that fiber laser beam quality indeed degrades due to the influence of transverse mode instability (TMI) effects [37], specifically manifesting as continuous energy coupling between multiple modes in output beams.

Addressing this problem, in 2017, Lincoln Laboratory achieved 1.2 kW stable output in a 25 μm core diameter ytterbium-doped fiber amplifier through active control of input beam phases in a 3 × 1 non-mode-selective photonic lantern [38], experimentally verifying that this scheme can improve system TMI threshold.

TMI is the result of thermally induced nonlinear effects [39]. Fundamental modes and higher-order modes propagate along active fibers with different phase velocities, forming periodic light intensity distributions (mode interference patterns, MIPs) during this process. This causes periodic thermal load distributions on active fibers, producing thermally induced refractive index changes (RIGs), ultimately inducing energy exchange between different modes in fibers.

The mechanism by which photonic lantern-based mode control systems suppress TMI is actively generating controllable mode superposition fields and using them as seeds for fiber laser main amplification stages. These mode superposition fields, after entering active fibers, can produce appropriate MIP to block TMI occurrence or terminate TMI occurrence. Blocking TMI occurrence means that during the process of system output power continuously increasing from below the inherent TMI threshold to exceeding the TMI threshold, using photonic lantern mode control systems to actively generate fast controllable mode superposition fields cancels out MIPs produced by defects and environmental disturbances in active fibers, preventing them from having time to produce thermal effects and preventing RIG formation. Terminating TMI refers to when fiber laser systems have already experienced TMI, generating MIPs conjugated to inherent MIPs so that their thermal effects "wash out" existing RIGs, ultimately achieving TMI termination.

Taking a 5 × 1 photonic lantern-based all-fiber mode control system as an example, its output beam is used as seed for a 42 μm core diameter ytterbium-doped fiber amplifier to verify its effect in blocking TMI occurrence. The gain fiber selected for the main amplification stage is NA = 0.075 non-polarization-maintaining fiber with absorption coefficient of approximately 18 dB/m @ 976 nm. When this amplifier uses conventional seed sources, the TMI power threshold P_{thr} is approximately 100 W, meaning that when output power exceeds 100 W, the time-domain signal dispersion of output beam power in bucket significantly increases (as shown in Figure 9a), with characteristic peaks beginning to appear around 800 Hz in the corresponding Fourier spectrum, and as power continues to increase, characteristic peaks show trends of narrowing width and increasing peak values (as shown in Figure 9b).

When using output beams from the photonic lantern all-fiber mode control system as seed sources: under closed-loop control algorithm conditions, continuously increasing pump power, when output power is below, above, or even four times P_{thr}, time-domain signal dispersion remains small (as shown by green, blue, and red curves in Figure 9c), with only occasional outliers appearing as power increases. Simultaneously, no characteristic peaks appear in corresponding Fourier spectrum plots (as shown by green, blue, and red curves in Figure 9d). If the control algorithm is turned off when output power is above P_{thr}, as shown by purple curves in Figure 9c,d, the time-domain signal interquartile range increases significantly with many outliers, characteristic peaks with low peak values and large widths appear in Fourier spectra (frequency coverage 0–600 Hz), average output power decreases significantly, gain fiber temperature rises rapidly, and sustained light output becomes impossible.

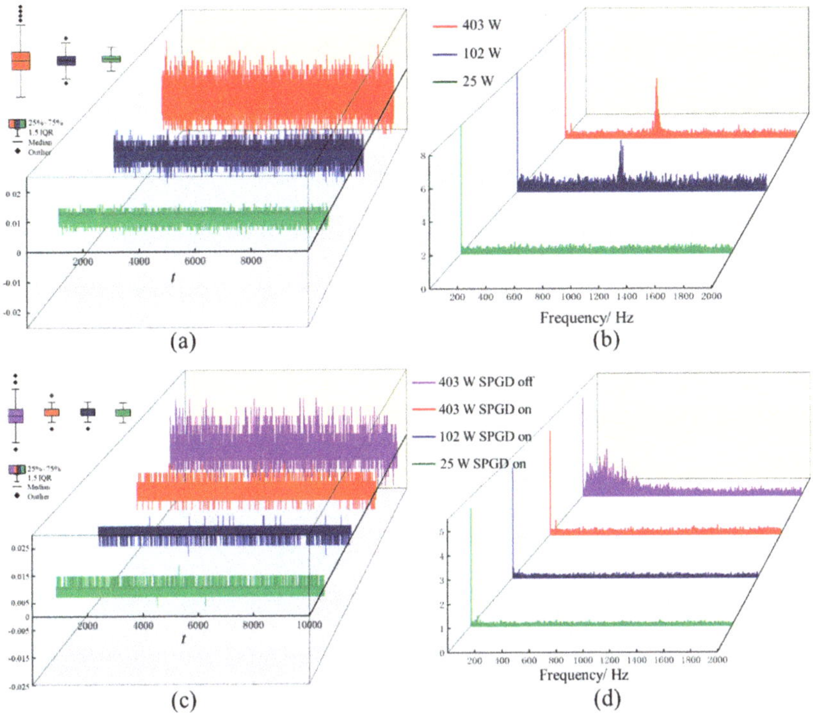

Figure 9. Changes in time-domain and spectral characteristics of output beams from ytterbium-doped fiber amplifiers with conventional/photonic lantern seed sources at different output powers [40]. (**a**,**c**) time-domain signal; (**b**,**d**) spectral signal.

Using M^2 analyzers to compare beam quality at P_{out} = 367 W, measurements show that when using conventional seed sources, due to TMI occurrence, output spots are lobed and continuously changing, with M_x^2 = 2.646 and M_y^2 = 2.503, indicating poor beam quality. When using photonic lantern mode control systems as seed sources under closed-loop system conditions, output spots exhibit stable quasi-fundamental mode forms with M_x^2 = 1.762 and M_y^2 = 1.567, indicating good beam quality. The above experimental results prove that photonic lantern mode control systems not only excellently suppress TMI but also significantly improve beam quality, well meeting demands for high-power, high-brightness, large mode field laser output in industrial manufacturing and other high-energy laser systems.

4.2. Optical Fiber Communications

To increase capacity, improve efficiency, and ensure security, optical communication system builders are dedicated to researching various multiplexing techniques or coding techniques to achieve multi-dimensional development and utilization of optical wave resources. Following time domain, wavelength, and polarization of light, the spatial distribution (mode) of light has become the newest (and possibly last) degree of freedom that can be utilized, considered an important handle for continued development of optical communication systems [41].

Photonic lanterns, as adiabatic mode converters, can evolve a group of discrete fundamental mode beams into a fundamental mode or higher-order mode with larger mode field area, and vice versa. This special light transmission characteristic ensures that they can excellently perform multiplexing/demultiplexing or coding/decoding functions in communication systems. Over the past decade, the practical value of photonic lanterns in multiple-input multiple-output (MIMO) networks and optical interconnection networks has been successively verified: In 2012, Nicolas K. Fontaine et al. proposed the concept of using three-channel mode-selective photonic lanterns for mode division multiplexing [42]. In 2014, Uden et al. [43] and Paul Mitchell et al. [44], respectively, designed 21 (3×7)-channel and 51 (3×19)-channel integrated photonic lanterns and used them in communication systems with three types of mode division multiplexing combined with seven (or nineteen) types of frequency multiplexing, not only improving communication capacity but also proving device reliability in multi-dimensional multiplexing communication technology. In 2021, A. Alarcón et al. from Sweden used photonic lanterns in quantum communication systems, achieving generation, encoding, and detection of three spatial photon states [45].

However, the photonic lanterns selected in the above optical communication systems are basically mode-selective photonic lanterns. The requirements for input fiber core diameter differences in structural design of this type of photonic lantern will undoubtedly become the greatest limiting factor for expanding the number of system mode multiplexable modes and an important source of inter-mode crosstalk. A feasible optimization method is to use non-mode-selective photonic lantern-based mode adaptive control systems for mode division multiplexing or encoding, then use the same photonic lanterns for mode division demultiplexing or decoding, as shown in Figure 8b.

When using photonic lantern all-fiber mode control systems to generate specific modes or mode superposition combinations, geometric arrangement schemes of single-mode fiber bundles can be optimized to improve control algorithm convergence speed. For LP modes supported by step-index fibers, their morphological patterns are as follows: for LP_{mn} modes, their spots have $2m$ lobes in the angular direction and $n - 1$ layers in the radial direction, and when $m > 0$, LP_{mn} modes have two degenerate states. Therefore, corresponding photonic lantern single-mode end fiber core arrangement schemes should be $2m + 1$ fibers uniformly distributed in regular polygons in the angular direction, with $n - 1$ layers of fibers distributed in the radial direction. Following the above method, Figure 10 shows geometric arrangements of single-mode fiber bundles for different channel number photonic lanterns and their corresponding LP modes.

When two degenerate states of LP_{mn} modes ($m > 0$) are linearly superposed with equal amplitudes and phase differences of $\pm\pi/2$, they can form orbital angular momentum (OAM) modes [46,47]. Their light field distribution $E(OAM_p{}^l)$ can be written as follows:

$$E(OAM_p{}^{\pm l}) = E(LP_{mno}) \pm e^{i\pi/2}E(LP_{mne}),\ l = m, p = n - 2 \tag{5}$$

where l is the topological charge number, and each photon in the beam carries average orbital angular momentum of $l\hbar$ ($\hbar$ is Planck's constant). Therefore, digital encoding can be performed according to different orbital angular momentum numbers carried by OAM modes. The phase expression of OAM modes contains the spiral term $e^{il\varphi}$, making different OAM modes highly orthogonal to each other, enabling good mode division multiplexing. Particularly, when control targets are OAM mode sequences, the central fiber in 6×1, 8×1, 15×1, ... photonic lantern single-mode fiber bundles can be replaced with no-core fibers to improve control efficiency [48].

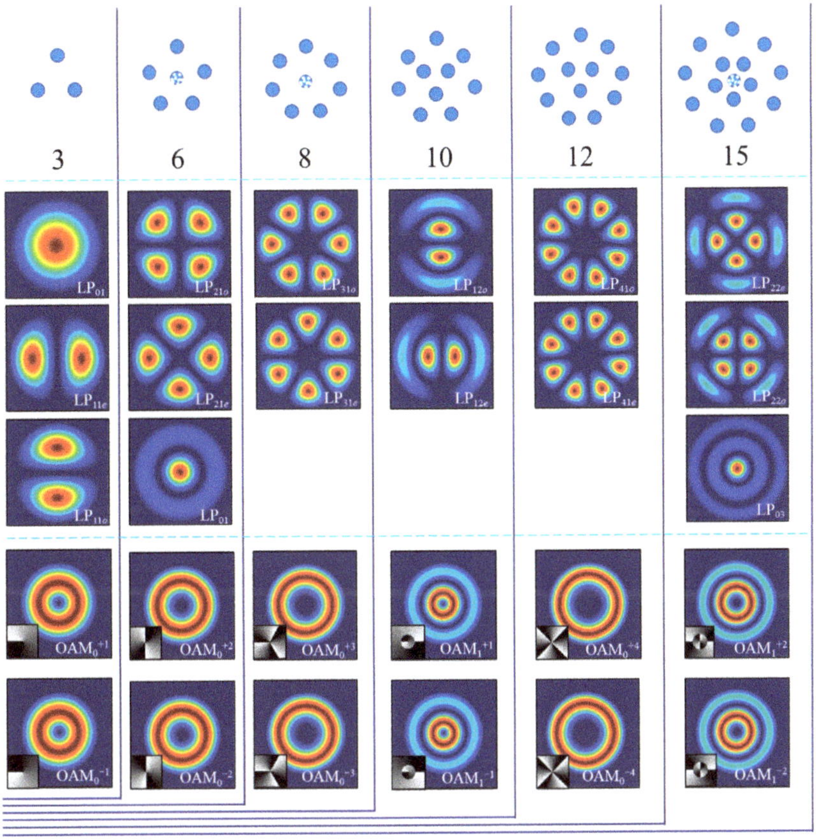

Figure 10. Single-mode fiber bundle arrangements for different channel number photonic lanterns and their mode control ranges.

When photonic lantern-based all-fiber mode control is used for optical fiber communications, the number of tunable modes depends on photonic lantern channel numbers. Currently reported photonic lantern channel numbers have reached several hundred [49], meaning theoretically hundreds of modes can be tuned. Mode switching speed depends on system control bandwidth. Optimization algorithm-based adaptive control systems can easily achieve control bandwidth above 10 kHz, so systems are expected to achieve mode switching speeds on the order of 10 kHz, promising to play irreplaceable roles in optical communication systems.

4.3. Frontier Exploration

Photonic lantern-based mode control systems possess advantages of all-fiber implementation, compact structure, low loss, and single-aperture fiber output, enabling applications in many frontier exploration fields.

In 2018, Velázquez-Benítez et al. employed three-channel mode-selective photonic lanterns to achieve single-particle trapping and manipulation. Additionally, they utilized the large-mode-field, switchable linearly-polarized modes output from photonic lanterns to realize the trapping and rotation of double-particles or particle clusters. This approach demonstrated extremely high flexibility at the single-aperture multimode fiber output end [50]. Furthermore, if photonic lantern-based all-fiber mode control systems are used to generate freely switchable higher-order linearly polarized modes carrying gradient forces or OAM modes carrying orbital angular momentum, simultaneous particle trapping, rotation, and transport in multiple planes could be achieved, as shown in Figure 8c.

In 2021, Moraitis et al. developed a novel photonic lantern approach for PolyOculus fiber optic linkages that potentially offers substantial advantages over previously considered free-space optical linkages, including much higher coupling efficiencies [51]. PolyOculus is for producing large-area-equivalent telescopes by using photonic technologies to link modules of multiple semi-autonomous, small, inexpensive, commercial-off-the-shelf telescopes. We can boldly imagine that in the future, introducing adaptive control systems and properly modulating the input ends of photonic lanterns will further improve throughput efficiency and beam quality, resulting in better astronomical observation results.

In 2024, Liang et al. proposed a high-precision all-fiber mode division multiplexing reconstructive spectrometer (RS) [52]. By using photonic lanterns to adjust mode field distribution conditions incident on scattering media, spectra are more comprehensively encoded into multiple speckle patterns. They achieved 2 pm spectral resolution and reconstruction of 2000 spectral channels with significantly reduced recovery errors, as shown in Figure 8d. In the future, by increasing photonic lantern channel numbers and thereby increasing generated mode types, efficient application to higher precision, higher resolution spectral measurement scenarios could be achieved.

5. Conclusions

In this paper, we have extensively reviewed the core mode evolution device, adaptive control mechanism, and applications of the transverse mode adaptive control system based on photonic lanterns. Regarding system cost, this all-fiber system employs internal optical path modulation, resulting in a compact system structure. In terms of process complexity, the fabrication difficulty of photonic lanterns is significantly lower compared to that of specialty optical fibers and advanced metasurface devices. In terms of power scalability, the photonic-lantern-based mode control system is an all-fiber system with single-aperture output. It can be efficiently and low-loss fused with common fiber laser systems (especially fiber laser amplification stages), exhibiting excellent power amplification capabilities. We look forward to the rise of new applications, as this system possesses advantages of all-fiber implementation, compact structure, low loss, and single-aperture fiber output.

Author Contributions: Conceptualization, Y.L. and Z.J.; methodology, Y.L. and Z.C.; software, Y.L. and T.L.; validation, Y.L. and Z.S.; resources, Y.L. and Z.C.; writing—original draft preparation, Y.L.; writing—review and editing, Z.S. and T.L. All authors have read and agreed to the published version of the manuscript.

Funding: This work was supported by National Natural Science Foundation of China (62405378).

Data Availability Statement: The original contributions presented in the study are included in the article; further inquiries can be directed to the corresponding authors.

Conflicts of Interest: The authors declare no conflicts of interest.

References

1. Shiner, B. Delivering power. *Nat. Photonics* **2010**, *4*, 290. [CrossRef]
2. Paterson, L.; MacDonald, M.P.; Arlt, J.; Sibbett, W.; Bryant, P.E.; Dholakia, K. Controlled Rotation of Optically Trapped Microscopic Particles. *Science* **2001**, *292*, 912–914. [CrossRef] [PubMed]
3. Li, L.; Gattass, R.R.; Gershgoren, E.; Hwang, H.; Fourkas, J.T. Achieving λ/20 Resolution by One-Color Initiation and Deactivation of Polymerization. *Science* **2009**, *324*, 910–913. [CrossRef] [PubMed]
4. Willner, A.E.; Huang, H.; Yan, Y.; Ren, Y.; Ahmed, N.; Xie, G.; Bao, C.; Li, L.; Cao, Y.; Zhao, Z.; et al. Optical communications using orbital angular momentum beams. *Adv. Opt. Photon.* **2015**, *7*, 66–106. [CrossRef]

5. Freeman, R.H.; Pearson, J.E. Deformable Mirrors for All Seasons and Reasons. *Appl. Opt.* **1982**, *21*, 580–588. [CrossRef]
6. Hayasaki, Y.; Sugimoto, T.; Takita, A.; Nishida, N. Variable holographic femtosecond laser processing by use of a spatial light modulator. *Appl. Phys. Lett.* **2005**, *87*, 031101. [CrossRef]
7. Bishara, W.; Sikora, U.; Mudanyali, O.; Su, T.-W.; Yaglidere, O.; Luckhart, S.; Ozcan, A. Holographic pixel super-resolution in portable lens less on-chip microscopy using a fiber-optic array. *Lab Chip* **2011**, *11*, 1276–1279. [CrossRef]
8. Geng, C.; Li, X.; Zhang, X.; Rao, C. Coherent beam combination of an optical array using adaptive fiber optics collimators. *Opt. Commun.* **2011**, *284*, 5531–5536. [CrossRef]
9. Su, R.; Yang, B.; Xi, X.; Zhou, P.; Wang, X.; Ma, Y.; Xu, X.; Chen, J. 500 W level MOPA laser with switchable output modes based on active control. *Opt. Express* **2017**, *25*, 23275–23282. [CrossRef]
10. Li, L.; Lou, Q.; Zhou, J.; Dong, J.; Wei, Y.; Du, S.; He, B. High power single transverse mode operation of a tapered large-mode-area fiber laser. *Opt. Commun.* **2008**, *281*, 655–657. [CrossRef]
11. Ma, X.; Zhu, C.; Hu, I.N.; Kaplan, A.; Galvanauskas, A. Single-mode chirally-coupled-core fibers with larger than 50μm diameter cores. *Opt. Express* **2014**, *22*, 9206–9219. [CrossRef]
12. Limpert, J.; Stutzki, F.; Jansen, F.; Otto, H.-J.; Eidam, T.; Jauregui, C.; Tünnermann, A. Yb-doped large-pitch fibres: Effective single-mode operation based on higher-order mode delocalization. *Light Sci. Appl.* **2012**, *1*, e8. [CrossRef]
13. Dong, L.; McKay, H.A.; Fu, L.; Ohta, M.; Marcinkevicius, A.; Suzuki, S.; Fermann, M.E. Ytterbium-doped all glass leakage channel fibers with highly fluorine-doped silica pump cladding. *Opt. Express* **2009**, *17*, 8962–8969. [CrossRef]
14. Chen, X.; Li, M.-J.; Venkataraman, N.; Gallagher, M.T.; Wood, W.A.; Crowley, A.M.; Carberry, J.P.; Zenteno, L.A.; Koch, K.W. Highly birefringent hollow-core photonic bandgap fiber. *Opt. Express* **2004**, *12*, 3888–3893. [CrossRef] [PubMed]
15. Jain, D.; Baskiotis, C.; Sahu, J.K. Mode area scaling with multi-trench rod-type fibers. *Opt. Express* **2013**, *21*, 1448–1455. [CrossRef]
16. Sun, B.; Wang, A.; Xu, L.; Gu, C.; Lin, Z.; Ming, H.; Zhan, Q. Low-threshold single-wavelength all-fiber laser generating cylindrical vector beams using a few-mode fiber Bragg grating. *Opt. Lett.* **2012**, *37*, 464–466. [CrossRef]
17. Tsekrekos, C.P.; Syvridis, D. Symmetric Few-Mode Fiber Couplers as the Key Component for Broadband Mode Multiplexing. *J. Light. Technol.* **2014**, *32*, 2461–2467. [CrossRef]
18. Kim, M.; Wong, A.M.H.; Eleftheriades, G.V. Optical Huygens' Metasurfaces with Independent Control of the Magnitude and Phase of the Local Reflection Coefficients. *Phys. Rev. X* **2014**, *4*, 041042. [CrossRef]
19. Montoya, J.; Aleshire, C.; Hwang, C.; Fontaine, N.K.; Velázquez-Benítez, A.; Martz, D.H.; Fan, T.Y.; Ripin, D. Photonic lantern adaptive spatial mode control in LMA fiber amplifiers. *Opt. Express* **2016**, *24*, 3405–3413. [CrossRef]
20. Birks, T.A.; Gris-Sánchez, I.; Yerolatsitis, S.; Leon-Saval, S.G.; Thomson, R.R. The photonic lantern. *Adv. Opt. Photon.* **2015**, *7*, 107–167. [CrossRef]
21. Bland-Hawthorn, J.; Leon-Saval, S.G. Astrophotonics: Molding the flow of light in astronomical instruments [Invited]. *Opt. Express* **2017**, *25*, 15549–15557. [CrossRef]
22. Leon-Saval, S.G.; Birks, T.A.; Bland-Hawthorn, J.; Englund, M. Multimode fiber devices with single-mode performance. *Opt. Lett.* **2005**, *30*, 2545–2547. [CrossRef]
23. Leon-Saval, S.G.; Argyros, A.; Bland-Hawthorn, J. Photonic lanterns: A study of light propagation in multimode to single-mode converters. *Opt. Express* **2010**, *18*, 8430–8439. [CrossRef]
24. Fontaine, N.K.; Ryf, R.; Bland-Hawthorn, J.; Leon-Saval, S.G. Geometric requirements for photonic lanterns in space division multiplexing. *Opt. Express* **2012**, *20*, 27123–27132. [CrossRef] [PubMed]
25. David, S.; Barnaby, R.M.N.; Peter, G.T.; Richard, S.; Jin, W.; Christopher, H.B.; Sergio, G.L.-S. Learning the lantern: Neural network applications to broadband photonic lantern modeling. *J. Astron. Telesc. Instrum. Syst.* **2021**, *7*, 028007.
26. Leon-Saval, S.G.; Argyros, A.; Bland-Hawthorn, J. Photonic lanterns. *Nanophotonics* **2013**, *2*, 429–440. [CrossRef]
27. Velazquez-Benitez, A.M.; Alvarado, J.C.; Lopez-Galmiche, G.; Antonio-Lopez, J.E.; Hernández-Cordero, J.; Sanchez-Mondragon, J.; Sillard, P.; Okonkwo, C.M.; Amezcua-Correa, R. Six mode selective fiber optic spatial multiplexer. *Opt. Lett.* **2015**, *40*, 1663–1666. [CrossRef]
28. Van Roey, J.; van der Donk, J.; Lagasse, P.E. Beam-propagation method: Analysis and assessment. *J. Opt. Soc. Am.* **1981**, *71*, 803–810. [CrossRef]
29. Sztefka, G.; Nolting, H.P. Bidirectional eigenmode propagation for large refractive index steps. *IEEE Photonics Technol. Lett.* **1993**, *5*, 554–557. [CrossRef]
30. Lu, Y.; Liu, W.; Chen, Z.; Jiang, M.; Zhou, Q.; Zhang, J.; Li, C.; Chai, J.; Jiang, Z. Spatial mode control based on photonic lanterns. *Opt. Express* **2021**, *29*, 41788–41797. [CrossRef]
31. Lu, Y.; Chen, Z.; Liu, W.; Jiang, M.; Yang, J.; Zhou, Q.; Zhang, J.; Chai, J.; Jiang, Z. Stable single transverse mode excitation in 50 um core fiber using a photonic-lantern-based adaptive control system. *Opt. Express* **2022**, *30*, 22435–22441. [CrossRef]
32. Noordegraaf, D.; Skovgaard, P.M.W.; Nielsen, M.D.; Bland-Hawthorn, J. Efficient multi-mode to single-mode coupling in a photonic lantern. *Opt. Express* **2009**, *17*, 1988–1994. [CrossRef]

33. Davenport, J.J.; Diab, M.; Deka, P.J.; Tripathi, A.; Madhav, K.; Roth, M.M. Photonic lanterns: A practical guide to filament tapering. *Opt. Mater. Express* **2021**, *11*, 2639–2649. [CrossRef]
34. Vorontsov, M.A.; Sivokon, V.P. Stochastic parallel-gradient-descent technique for high-resolution wave-front phase-distortion correction. *J. Opt. Soc. Am. A* **1998**, *15*, 2745–2758. [CrossRef]
35. Beck, T.; Reng, N.; Weber, H. Optical fibres for material processing lasers. *Opt. Lasers Eng.* **2000**, *34*, 255–272. [CrossRef]
36. Dawson, J.W.; Messerly, M.J.; Beach, R.J.; Shverdin, M.Y.; Stappaerts, E.A.; Sridharan, A.K.; Pax, P.H.; Heebner, J.E.; Siders, C.W.; Barty, C.P.J. Analysis of the scalability of diffraction-limited fiber lasers and amplifiers to high average power. *Opt. Express* **2008**, *16*, 13240–13266. [CrossRef] [PubMed]
37. Eidam, T.; Wirth, C.; Jauregui, C.; Stutzki, F.; Jansen, F.; Otto, H.-J.; Schmidt, O.; Schreiber, T.; Limpert, J.; Tünnermann, A. Experimental observations of the threshold-like onset of mode instabilities in high power fiber amplifiers. *Opt. Express* **2011**, *19*, 13218–13224. [CrossRef] [PubMed]
38. Montoya, J.; Hwang, C.; Martz, D.; Aleshire, C.; Fan, T.Y.; Ripin, D.J. Photonic lantern kW-class fiber amplifier. *Opt. Express* **2017**, *25*, 27543–27550. [CrossRef]
39. Kong, F.; Xue, J.; Stolen, R.H.; Dong, L. Direct experimental observation of stimulated thermal Rayleigh scattering with polarization modes in a fiber amplifier. *Optica* **2016**, *3*, 975–978. [CrossRef]
40. Jiang, Z.; Lu, Y.; Liu, W.; Zhou, Q.; Jiang, M.; Chen, Z.; Zhang, H.; Liu, P.; Zhang, J. All-Fiber Spatial Mode Excitation and Adaptive Control Based on Photonic Lanterns. *Acta Opt. Sin.* **2023**, *43*, 1700002.
41. Winzer, P.J. Making spatial multiplexing a reality. *Nat. Photonics* **2014**, *8*, 345–348. [CrossRef]
42. Fontaine, N.K.; Ryf, R.R.; Leon-Saval, S.G.; Bland-Hawthorn, J. Evaluation of photonic lanterns for lossless mode-multiplexing. In Proceedings of the 2012 38th European Conference and Exhibition on Optical Communications, Amsterdam, The Netherlands, 16–20 September 2012; pp. 1–3.
43. van Uden, R.G.H.; Correa, R.A.; Lopez, E.A.; Huijskens, F.M.; Xia, C.; Li, G.; Schülzgen, A.; de Waardt, H.; Koonen, A.M.J.; Okonkwo, C.M. Ultra-high-density spatial division multiplexing with a few-mode multicore fibre. *Nat. Photonics* **2014**, *8*, 865–870. [CrossRef]
44. Mitchell, P.; Brown, G.; Thomson, R.; Psaila, N.; Kar, A. 57 channel (19×3) spatial multiplexer fabricated using direct laser inscription. In Proceedings of the OFC 2014, San Francisco, CA, USA, 9–13 March 2014; pp. 1–3.
45. Alarcón, A.; Argillander, J.; Spegel-Lexne, D.; Xavier, G.B. Dynamic generation of photonic spatial quantum states with an all-fiber platform. *Opt. Express* **2023**, *31*, 10673–10683. [CrossRef]
46. Yao, A.M.; Padgett, M.J. Orbital angular momentum: Origins, behavior and applications. *Adv. Opt. Photon.* **2011**, *3*, 161–204. [CrossRef]
47. Forbes, A.; de Oliveira, M.; Dennis, M.R. Structured light. *Nat. Photonics* **2021**, *15*, 253–262. [CrossRef]
48. Lu, Y.; Jiang, Z.; Chen, Z.; Jiang, M.; Yang, J.; Zhou, Q.; Zhang, J.; Zhang, D.; Chai, J.; Yang, H.; et al. High-Power Orbital Angular Momentum Beam Generation Using Adaptive Control System Based on Mode Selective Photonic Lantern. *J. Light. Technol.* **2023**, *41*, 5607–5613. [CrossRef]
49. Birks, T.A.; Mangan, B.J.; Díez, A.; Cruz, J.L.; Murphy, D.F. "Photonic lantern" spectral filters in multi-core fibre. *Opt. Express* **2012**, *20*, 13996–14008. [CrossRef]
50. Velázquez-Benítez, A.M.; Guerra-Santillán, K.Y.; Caudillo-Viurquez, R.; Antonio-López, J.E.; Amezcua-Correa, R.; Hernández-Cordero, J. Optical trapping and micromanipulation with a photonic lantern-mode multiplexer. *Opt. Lett.* **2018**, *43*, 1303–1306. [CrossRef] [PubMed]
51. Moraitis, C.D.; Alvarado-Zacarias, J.C.; Amezcua-Correa, R.; Jeram, S.; Eikenberry, S.S. Demonstration of high-efficiency photonic lantern couplers for PolyOculus. *Appl. Opt.* **2021**, *60*, D93–D99. [CrossRef]
52. Liang, J.; Ye, J.; Ma, X.; Lu, Y.; Li, J.; Xu, J.; Chen, Z.; Leng, J.; Jiang, Z.; Zhou, P. Mode division multiplexing reconstructive spectrometer with an all-fiber photonics lantern. *Front. Optoelectron.* **2024**, *17*, 23. [CrossRef] [PubMed]

MDPI AG
Grosspeteranlage 5
4052 Basel
Switzerland
Tel.: +41 61 683 77 34

Micromachines Editorial Office
E-mail: micromachines@mdpi.com
www.mdpi.com/journal/micromachines